5th edition

# geog.3

## teacher's handbook

<rosemarie gallagher> <katy patchwood> <catherine hurst>

OXFORD
UNIVERSITY PRESS

OXFORD
UNIVERSITY PRESS

Great Clarendon Street, Oxford OX2 6DP

Oxford University Press is a department of the University of Oxford. It furthers the University's objective of excellence in research, scholarship, and education by publishing worldwide. Oxford is a registered trade mark of Oxford University Press in the UK and in certain other countries



Authors: RoseMarie Gallagher, Katy Patchwood, Catherine Hurst



First published 2002
Second Edition 2005
Third Edition 2009
Fourth Edition 2016
Fifth Edition 2023



British Library Cataloguing in Publication Data

Data available

ISBN: 978-0-19-848995-5

10 9 8 7 6 5 4 3 2 1

Paper used in the production of this book is a natural, recyclable product made from wood grown in sustainable forests.

The manufacturing process conforms to the environmental regulations of the country of origin.

Printed in Great Britain by CPI Group (UK) Ltd., Croydon CR0 4YY

**Acknowledgements**

The publisher would like to thank the following for permission to use photographs and other copyright material:

Cover image by: Shutterstock/OUP

Although we have made every effort to trace and contact all copyright holders before publication this has not been possible in all cases. If notified, the publisher will rectify any errors or omissions at the earliest opportunity.

Links to third party websites are provided by Oxford in good faith and for information only. Oxford disclaims any responsibility for the materials contained in any third party website referenced in this work.

# Contents

# About this course

*geog.3* is the second book of *geog.123* – the complete geography course for KS3. The course covers the KS3 Programme of Study, offers excellent support for assessment, and provides effective preparation for GCSE study.

## The course components

The course consists of:

### For students

- three students' books – also available as Kerboodle online books
- three workbooks

### For teachers

- three handbooks
- three workbook answer books
- three Kerboodles of online lessons, resources, and assessment, including Kerboodle online teacher's handbooks and students' books for teacher access

Find out more about the course components by looking at these panels.

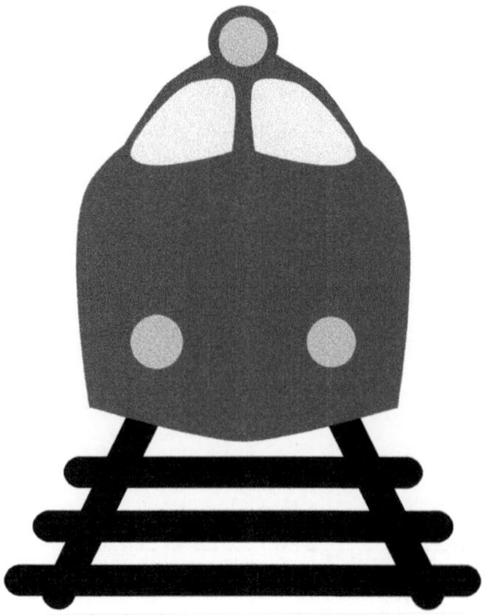

### The students' books

- Three books for the course
- Chapters divided into two-page units
- Aims of unit given in student-friendly language at the start of each unit
- 'Your turn' questions at the end of each unit
- 'Check' questions at the end of each chapter check knowledge, understanding, and skills across the chapter, with an opportunity for extended writing

### The workbooks

- One for each students' book
- Support for each unit in the students' book
- Activities support the development of key skills and knowledge of key terms
- Fill-in activities
- Ideal for homework and independent study
- Answers provided in separate answer books

### The teacher's handbooks

- One for each students' book
- Chapter overviews
- Help at a glance for each unit
- Ideas for starters and plenaries for each unit
- Outcomes for each unit
- Answers for 'Your turn' and 'Check' questions
- Further suggestions for class and homework
- Glossary

*geog.123* provides a wide range of materials. The students' books are the core of the course. They combine a rigorous approach to content with a uniquely engaging style.

You can decide how to use the wealth of support materials, but notes in the teacher's handbooks will point you towards appropriate material in the workbooks and *Kerboodle Lessons, Resources, & Assessment*. The result is a truly comprehensive and flexible geography course – which we hope you will enjoy using.

kerboodle

***Kerboodle Lessons, Resources, & Assessment***

- A package to support each students' book
- Front-of-class lesson activities and teacher notes for each chapter
- Interactive activities – for front-of-class use or homework
- Animations and supporting worksheets, with answers
- Skills-pods and supporting worksheets, with answers
- Glossaries and glossary worksheets – one for each chapter
- Editable and photocopiable differentiated Foundation and Extension worksheets for each unit, with answers provided
- Two Week Topics – project-based independent-learning resources for students to complete with minimal supervision
- Teacher access to the Kerboodle Book – the online version of the students' book – for front-of-class use, with links to resources
- Teacher access to an online version of the teacher's handbook, with links to resources
- Comprehensive support for assessment:
  - interactive auto-marked End-of-lesson tests, with feedback – one for each unit
  - interactive auto-marked chapter tests
  - End-of-chapter assessment papers, with mark schemes and reporting framework
  - editable and photocopiable Self-assessment checklists – one for each chapter
  - markbook and reporting functions
- Many resources are editable
- Upload and create your own content

For more information, see pages 12–17.

kerboodle

You can access *geog.123* online. There are two purchasable options:

- *Kerboodle Lessons, Resources, & Assessment* (which includes teacher access to the *Kerboodle Book*, an online version of the students' book)
- student access to the *Kerboodle Book*

You can choose to use one or the other, or both – you decide what you and your students need.

kerboodle

***Kerboodle Book***

- An online version of each of the students' books
- Includes a range of tools that allow students to annotate their book
- Can be accessed on a range of devices with Internet connectivity, including iPads and other tablets – so students can have access anywhere, any time

For more information, see page 13.

The 'Teaching versions' of the students' book and teacher's handbook feature clickable 'hotspots' that link directly to relevant Kerboodle resources in the Lessons and Resources tabs – to help you find things quickly and easily.

You can assign assessments to students from the Assessment tab, and then make use of the markbook and reporting functionality. The interactive tests could also be used front-of-class.

The lists of resources provided in this book are up-to-date at the time of writing. However, Kerboodle is a dynamic product and is constantly evolving and improving. Please explore the platform and take note of the new content alerts on the homepage to get the most out of your Kerboodle.

# Using this book

This book aims to save you time and effort! It offers full support for the *geog.3* students' book, and will help you prepare detailed course and lesson plans.

## What it provides

For each chapter of the students' book, this book provides:

**1** a chapter overview

**2** help at a glance for each unit, including answers for 'Your turn' activities

**3** answers for the end-of-chapter 'Check' questions

**4** further suggestions for class and homework.

It also has a glossary at the back, covering the geographical terms the students will meet.

Please turn to the contents list on page 3 now, to see how this book is structured. Then find out more about the four main components, below.

## 1 The chapter overview

This is your introduction to the corresponding students' chapter. Look at its sections.

Shows how the students' chapter relates to the KS3 Programme of Study.

Sets out the objectives and outcomes for the chapter, and the corresponding unit numbers.

Sets out the key ideas within, and behind, the students' chapter.

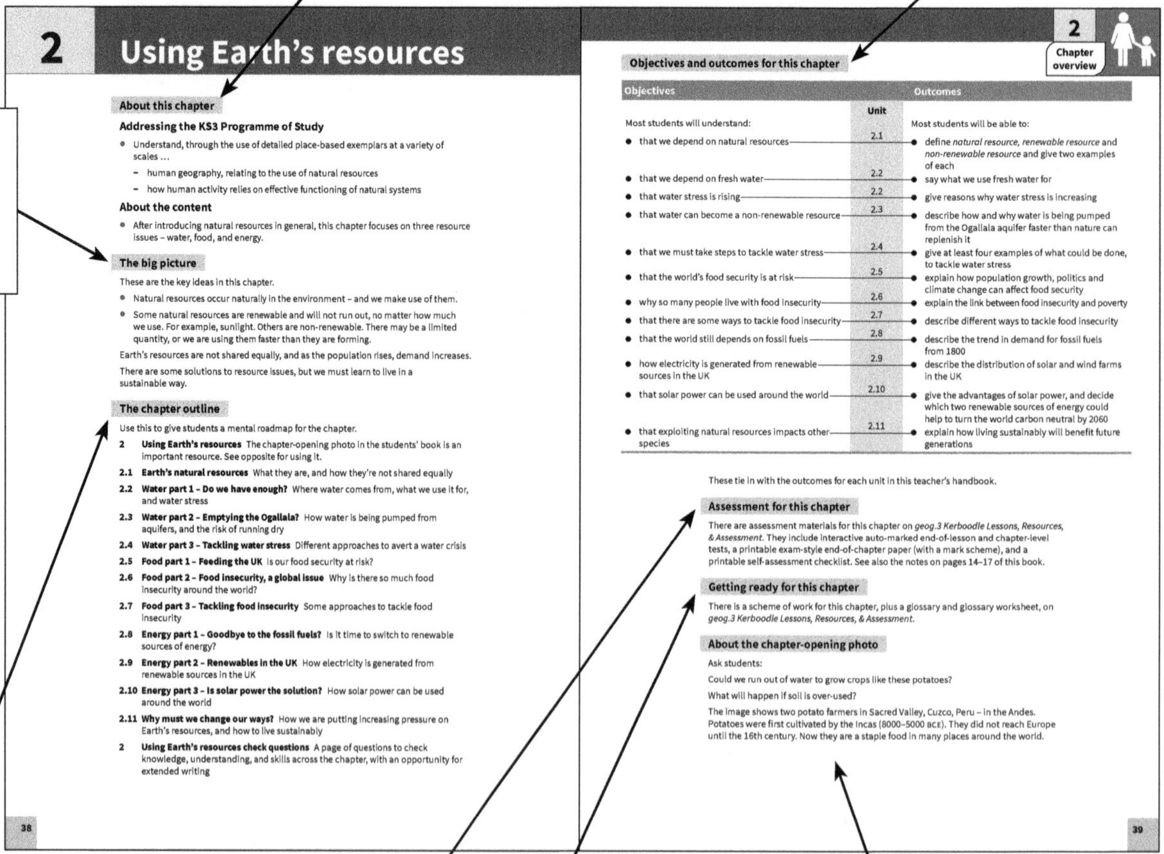

2 Using Earth's resources

**About this chapter**

**Addressing the KS3 Programme of Study**

- Understand, through the use of detailed place-based exemplars at a variety of scales …
  - human geography, relating to the use of natural resources
  - how human activity relies on effective functioning of natural systems

**About the content**

- After introducing natural resources in general, this chapter focuses on three resource issues – water, food, and energy.

**The big picture**

These are the key ideas in this chapter.

- Natural resources occur naturally in the environment – and we make use of them.
- Some natural resources are renewable and will not run out, no matter how much we use. For example, sunlight. Others are non-renewable. There may be a limited quantity, or we are using them faster than they are forming.

Earth's resources are not shared equally, and as the population rises, demand increases.

There are some solutions to resource issues, but we must learn to live in a sustainable way.

**The chapter outline**

Use this to give students a mental roadmap for the chapter.

2 **Using Earth's resources** The chapter-opening photo in the students' book is an important resource. See opposite for using it.

2.1 **Earth's natural resources** What they are, and how they're not shared equally

2.2 **Water part 1 – Do we have enough?** Where water comes from, what we use it for, and water stress

2.3 **Water part 2 – Emptying the Ogallala?** How water is being pumped from aquifers, and the risk of running dry

2.4 **Water part 3 – Tackling water stress** Different approaches to avert a water crisis

2.5 **Food part 1 – Feeding the UK** Is our food security at risk?

2.6 **Food part 2 – Food insecurity, a global issue** Why is there so much food insecurity around the world?

2.7 **Food part 3 – Tackling food insecurity** Some approaches to tackle food insecurity

2.8 **Energy part 1 – Goodbye to the fossil fuels?** Is it time to switch to renewable sources of energy?

2.9 **Energy part 2 – Renewables in the UK** How electricity is generated from renewable sources in the UK

2.10 **Energy part 3 – Is solar power the solution?** How solar power can be used around the world

2.11 **Why must we change our ways?** How we are putting increasing pressure on Earth's resources, and how to live sustainably

2 **Using Earth's resources check questions** A page of questions to check knowledge, understanding, and skills across the chapter, with an opportunity for extended writing

38

2 Chapter overview

**Objectives and outcomes for this chapter**

| Objectives | Unit | Outcomes |
|---|---|---|
| Most students will understand: | | Most students will be able to: |
| • that we depend on natural resources | 2.1 | • define *natural resource, renewable resource* and *non-renewable resource* and give two examples of each |
| • that we depend on fresh water | 2.2 | • say what we use fresh water for |
| • that water stress is rising | 2.2 | • give reasons why water stress is increasing |
| • that water can become a non-renewable resource | 2.3 | • describe how and why water is being pumped from the Ogallala aquifer faster than nature can replenish it |
| • that we must take steps to tackle water stress | 2.4 | • give at least four examples of what could be done, to tackle water stress |
| • that the world's food security is at risk | 2.5 | • explain how population growth, politics and climate change can affect food security |
| • why so many people live with food insecurity | 2.6 | • explain the link between food insecurity and poverty |
| • that there are some ways to tackle food insecurity | 2.7 | • describe different ways to tackle food insecurity |
| • that the world still depends on fossil fuels | 2.8 | • describe the trend in demand for fossil fuels from 1800 |
| • how electricity is generated from renewable sources in the UK | 2.9 | • describe the distribution of solar and wind farms in the UK |
| • that solar power can be used around the world | 2.10 | • give the advantages of solar power, and decide which two renewable sources of energy could help to turn the world carbon neutral by 2060 |
| • that exploiting natural resources impacts other species | 2.11 | • explain how living sustainably will benefit future generations |

These tie in with the outcomes for each unit in this teacher's handbook.

**Assessment for this chapter**

There are assessment materials for this chapter on *geog.3 Kerboodle Lessons, Resources, & Assessment*. They include interactive auto-marked end-of-lesson and chapter-level tests, a printable exam-style end-of-chapter paper (with a mark scheme), and a printable self-assessment checklist. See also the notes on pages 14–17 of this book.

**Getting ready for this chapter**

There is a scheme of work for this chapter, plus a glossary and glossary worksheet, on *geog.3 Kerboodle Lessons, Resources, & Assessment*.

**About the chapter-opening photo**

Ask students:

Could we run out of water to grow crops like these potatoes?

What will happen if soil is over-used?

The image shows two potato farmers in Sacred Valley, Cuzco, Peru – in the Andes. Potatoes were first cultivated by the Incas (8000–5000 BCE). They did not reach Europe until the 16th century. Now they are a staple food in many places around the world.

39

Gives a very brief summary of what's covered in the students' chapter. It will help you give students a roadmap for the chapter.

Points you to the assessment material for the chapter (written and electronic).

Reminds you that *geog.3 Kerboodle Lessons, Resources, & Assessment* contains valuable resource material.

Helps you make the most of the chapter-opening photo – questions you could ask students, and information about the photo.

## 2 Help at a glance for each unit

These pages give comprehensive help for each unit of the *geog.3* students' book.

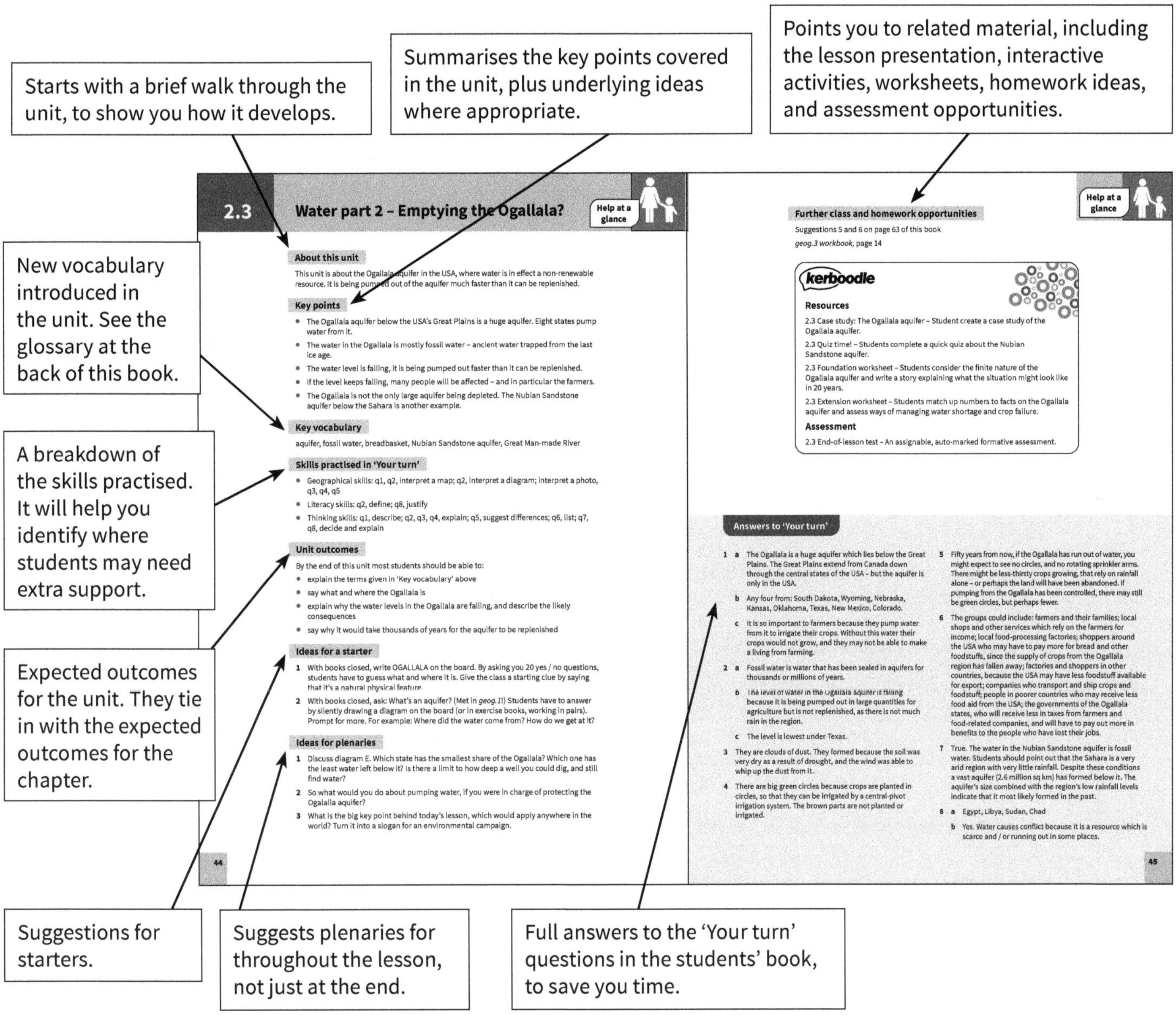


2.3 Water part 2 – Emptying the Ogallala?

Help at a glance

**About this unit**

This unit is about the Ogallala aquifer in the USA, where water is in effect a non-renewable resource. It is being pumped out of the aquifer much faster than it can be replenished.

**Key points**

- The Ogallala aquifer below the USA's Great Plains is a huge aquifer. Eight states pump water from it.
- The water in the Ogallala is mostly fossil water – ancient water trapped from the last ice age.
- The water level is falling, it is being pumped out faster than it can be replenished.
- If the level keeps falling, many people will be affected – and in particular the farmers.
- The Ogallala is not the only large aquifer being depleted. The Nubian Sandstone aquifer below the Sahara is another example.

**Key vocabulary**

aquifer, fossil water, breadbasket, Nubian Sandstone aquifer, Great Man-made River

**Skills practised in 'Your turn'**

- Geographical skills: q1, q2, interpret a map; q2, interpret a diagram; interpret a photo, q3, q4, q5
- Literacy skills: q2, define; q8, justify
- Thinking skills: q1, describe; q2, q3, q4, explain; q5, suggest differences; q6, list; q7, q8, decide and explain

**Unit outcomes**

By the end of this unit most students should be able to:

- explain the terms given in 'Key vocabulary' above
- say what and where the Ogallala is
- explain why the water levels in the Ogallala are falling, and describe the likely consequences
- say why it would take thousands of years for the aquifer to be replenished

**Ideas for a starter**

1 With books closed, write OGALLALA on the board. By asking you 20 yes / no questions, students have to guess what and where it is. Give the class a starting clue by saying that it's a natural physical feature.

2 With books closed, ask: What's an aquifer? (Met in *geog.1*!) Students have to answer by silently drawing a diagram on the board (or in exercise books, working in pairs). Prompt for more. For example: Where did the water come from? How do we get at it?

**Ideas for plenaries**

1 Discuss diagram E. Which state has the smallest share of the Ogallala? Which one has the least water left below it? Is there a limit to how deep a well you could dig, and still find water?

2 So what would you do about pumping water, if you were in charge of protecting the Ogalalla aquifer?

3 What is the big key point behind today's lesson, which would apply anywhere in the world? Turn it into a slogan for an environmental campaign.

44

Help at a glance

**Further class and homework opportunities**

Suggestions 5 and 6 on page 63 of this book

*geog.3 workbook*, page 14

kerboodle

**Resources**

2.3 Case study: The Ogallala aquifer – Student create a case study of the Ogallala aquifer.

2.3 Quiz time! – Students complete a quick quiz about the Nubian Sandstone aquifer.

2.3 Foundation worksheet – Students consider the finite nature of the Ogallala aquifer and write a story explaining what the situation might look like in 20 years.

2.3 Extension worksheet – Students match up numbers to facts on the Ogallala aquifer and assess ways of managing water shortage and crop failure.

**Assessment**

2.3 End-of-lesson test – An assignable, auto-marked formative assessment.

**Answers to 'Your turn'**

1 a The Ogallala is a huge aquifer which lies below the Great Plains. The Great Plains extend from Canada down through the central states of the USA – but the aquifer is only in the USA.

b Any four from: South Dakota, Wyoming, Nebraska, Kansas, Oklahoma, Texas, New Mexico, Colorado.

c It is so important to farmers because they pump water from it to irrigate their crops. Without this water their crops would not grow, and they may not be able to make a living from farming.

2 a Fossil water is water that has been sealed in aquifers for thousands or millions of years.

b The level of water in the Ogallala aquifer is falling because it is being pumped out in large quantities for agriculture but is not replenished, as there is not much rain in the region.

c The level is lowest under Texas.

3 They are clouds of dust. They formed because the soil was very dry as a result of drought, and the wind was able to whip up the dust from it.

4 There are big green circles because crops are planted in circles, so that they can be irrigated by a central-pivot irrigation system. The brown parts are not planted or irrigated.

5 Fifty years from now, if the Ogallala has run out of water, you might expect to see no circles, and no rotating sprinkler arms. There might be less-thirsty crops growing, that rely on rainfall alone – or perhaps the land will have been abandoned. If pumping from the Ogallala has been controlled, there may still be green circles, but perhaps fewer.

6 The groups could include: farmers and their families; local shops and other services which rely on the farmers for income; local food-processing factories; shoppers around the USA who may have to pay more for bread and other foodstuffs, since the supply of crops from the Ogallala region has fallen away; factories and shoppers in other countries, because the USA may have less foodstuff available for export; companies who transport and ship crops and foodstuff; people in poorer countries who may receive less food aid from the USA; the governments of the Ogallala states, who will receive less in taxes from farmers and food-related companies, and will have to pay out more in benefits to the people who have lost their jobs.

7 True. The water in the Nubian Sandstone aquifer is fossil water. Students should point out that the Sahara is a very arid region with very little rainfall. Despite these conditions a vast aquifer (2.6 million sq km) has formed below it. The aquifer's size combined with the region's low rainfall levels indicate that it most likely formed in the past.

8 a Egypt, Libya, Sudan, Chad

b Yes. Water causes conflict because it is a resource which is scarce and / or running out in some places.

45

## 3 Check questions: skills practised and answers

Full answers to the 'Check' questions in the students' book, to save you time. These questions check knowledge, understanding, and skills across each chapter, and provide an opportunity for extended writing.

## 4 Further suggestions for class and homework

These pages give a wealth of further suggestions for class and homework.

They have been graded *, ** or *** according to level of difficulty.

Some are suitable for all levels, and differentiated by outcome.

# Planning your lessons around *geog.3*

## Planning for high-quality lessons

Well-planned and well-structured lessons are a key requirement, for delivering high-quality teaching and learning in any subject, at any level. The *geog.123* course aims to make it easy to plan, structure, and deliver high-quality lessons for KS3 geography.

## Structure of a typical lesson

You will already be familiar with guidelines on structuring lessons. This shows a typical lesson structure.

### STARTER

*Purpose:* To capture students' attention and focus the class. Use it as the lesson hook, or to find out what students know already about a new topic, or for quick revision of earlier work.

### INTRODUCTION

*Purpose:* To prepare students for the activities ahead.

- If this is a new topic, tell students the topic objectives. Write these on the board.
- If it's a continuation of a topic, you can refer back to an objective as appropriate.

### ACTIVITIES

This is the main body of the lesson.

*Purpose:* To achieve one or more of the topic objectives.

- Emphasis on exploration and investigation.
- Provide for practice in different types of skill: geographical, literacy, numeracy, thinking, listening, speaking, team-working, and ICT skills.
- Choose from a variety of activities: reading, answering questions, enquiries, role play, game playing, fieldwork, and ICT.

**Plenaries:** note that plenaries can be used as staging posts throughout the activities, to gain feedback, check understanding, link to earlier work, and encourage reflection on what is being learned, and how.

### FINAL PLENARY

*Purpose:* To round off and review what has been done, and to assess what has been achieved against the topic objectives. This is where you help students to:

- check, and crystallise, their understanding
- generalise, for example from an individual case study
- set work in context, and make links to work already done, or to be done in the future
- reflect on how they have learned, as well as what
- check how well they have achieved the topic objectives (self-assessment).

### HOMEWORK

*Purpose:* To confirm, give practice in, and extend, what has been learned in the lesson.

- The homework can lead on from the final plenary, and be the basis for a starter for the next lesson.

## Planning around *geog.3*

Now see how the components of *geog.3* provide material for each part of your lesson.

### STARTERS

- The 'Help at a glance' pages in this book have suggestions for lesson starters.
- See further notes about starters, and resources for them, in this book.

### OBJECTIVES

- The opening lines of each unit in the students' book give the purpose of the unit, in student-friendly language.
- See also the objectives and outcomes given in this book.

### ACTIVITIES

**Using the students' book**

- The text in the students' book provides the core information students need. Some lends itself to reading aloud, but try 'quiet time' too.
- You can let students work through the text uninterrupted, or break it up with 'Your turn' questions. (These generally follow the order of the text.)
- The questions give practice in literacy, numeracy, thinking, and geography skills.
- Some are ideal as whole-class questions with verbal response. Others can be worked through by students working alone, in pairs, or in small groups. The final 'Your turn' questions are usually open questions that challenge students to show what they can do.
- For students who finish early, check out 'Further suggestions for class and homework' at the end of each chapter in this book. Or select a worksheet from the *Kerboodle Lessons, Resources, & Assessment*.

**Using *Kerboodle Lessons, Resources, & Assessment***

- The *Kerboodle Lessons, Resources, & Assessment* gives you ready-to-use front-of-class lesson activities (with teacher's notes) for each unit plus interactive activities, animations and 'skills-pods'.
- The *Kerboodle* also offers differentiated Foundation and Extension worksheets and an End-of-lesson interactive test (with feedback) for each unit.
- Much of the material is suitable for whole-class teaching using an interactive whiteboard or projector, and offers scope for vibrant and effective lessons.

### PLENARIES

- The 'Help at a glance' pages in this book give suggestions for plenaries, for throughout the lesson as well as at the end.
- See further notes about plenaries, and resources for them, in this book.

### HOMEWORK

- 'Further suggestions for class and homework' at the end of each chapter in this book offer lots of ideas.
- Select a worksheet, or an animation or 'skills-pod', from the *Kerboodle Lessons, Resources, & Assessment*.
- The workbook provides support for every unit in the students' book.
- Use the End-of-lesson interactive test for each unit or the End-of-chapter interactive test on the *Kerboodle Lessons, Resources, & Assessment*.
- Use other assessment materials on the *Kerboodle Lessons, Resources, & Assessment*.

# More about starters and plenaries

## Planning your starters and plenaries

Effective starters and plenaries need to be planned for. With planning, you can ensure that they'll help you to meet your lesson objectives, and that you won't have to rely on sudden inspiration in the classroom. But even where they are planned, you may want and need to modify them as you go along, in response to your students.

## Our suggestions for starters and plenaries

The kinds of activities you feel comfortable with, for starters and plenaries, will depend on your teaching style, and the individual class. So the suggestions for starters and plenaries in this book are just that: suggestions! You may want to use some as described, or adapt them. Or they may provide inspiration for new ideas of your own.

### The starters

- Most of these are intended for use with the students' books closed, before students have looked at the new unit. But they lead seamlessly into the work in the students' book.
- In some cases you may want to combine two starters to give a more extended one.
- A number of starters require the use of an atlas, and can be an excellent way of giving your students atlas practice that's fun.
- Other starters require both physical and mental activity – for example creating a graffiti wall on the board. This is a good way to get everyone involved.

### The plenaries

- There are suggestions for plenaries for throughout the lesson, not just at the end.
- They have been chosen for a variety of purposes: to encourage feedback; assess understanding; promote reflection; build bridges with material already covered (or still to be covered), with other subjects, and with the real world; help crystallise what has been learned; and see whether it applies to other situations.
- Some of the plenaries are single questions. You will find that you can readily combine some to make more extended plenaries.
- Some need more preparation than others. You might not want to choose these for every class, but it's a good idea to ring the changes, and keep your students surprised.
- Together with 'Your turn', the 'Ideas for plenaries' section is a rich resource to help you deliver fresh, exciting, and effective lessons.

## Resources for starters and plenaries

### Images

Many of the starters, and some plenaries, require images – mostly photos. There's a wealth of useful resources on the *Kerboodle Lessons, Resources, & Assessment*. You could use some of these to design your own starters or plenaries.

The interactive activities and animations and video clips on the *Kerboodle Lessons, Resources, & Assessment* also make great starters and plenaries.

The Internet, of course, is an excellent source for other geographical photos and images. Please check with the appropriate people in your school regarding copyright issues.

### Building a resource library

Some resources, such as photos, can be used over and over. You may want to create your own resource library. Laminating printed photos, and other resources (such as True / False cards), will extend their lives and save you time and effort in the future.

# Using the chapter openers in the students' book

The chapter openers in the students' book are in effect the starters for new topics – and you can return to them as an end-of-topic plenary.

Below is a typical chapter opener.

Most chapter openers are large photos – to hook your students' attention (we hope!). The opening photos usually relate to specific material within the chapter, and are referred back to, at different points.

The 'Chapter overview' pages in this book provide information about the chapter-opening photos (where available), and some background information – and question ideas to help you get your students thinking.

Using the teacher-access *Kerboodle Book* in *Kerboodle Lessons, Resources, & Assessment*, you could display the chapter openers on the whiteboard.

So, as you can see, the chapter openers can do quite a lot of useful work, so it's worth spending some time on them.

The 'Chapter overview' pages in this book detail the big underlying ideas for the chapter. These ideas provide the context for new learning. As you start the topic, 'The big picture' can be read aloud, and discussed. At the end of the topic, the big ideas can be reviewed, to help crystallise the learning.

Using the chapter-opening photos in the students' book, along with the questions and the big ideas for the topic provided in the 'Chapter overviews', we hope you will be able to introduce the topic, engage your students, and give them a mental roadmap for the chapter.

# About *geog.3 Kerboodle*

Kerboodle is an online resource. It offers two purchasable options:

- Student access to the *Kerboodle Book* – an online version of the students' book
- *Kerboodle Lessons, Resources, & Assessment* (which includes teacher access to a digital version of this teacher's handbook and to the *Kerboodle Book*)

You can choose to use one or the other, or both – you decide what you and your students need. Each option is available as an annual licence for unlimited users.

## *Kerboodle Book*

The *Kerboodle Book* is an online version of the students' book which includes a range of annotation tools, such as sticky notes and drawing tools. A 'Teaching version' of the students' book, with clickable hotspots that link directly to relevant Kerboodle resources, is included in the *Kerboodle Lessons, Resources, & Assessment.*

Student access includes a range of tools that allow students to annotate their book.

Teacher access is included as part of the *Kerboodle Lessons, Resources, & Assessment* for front-of-class use.

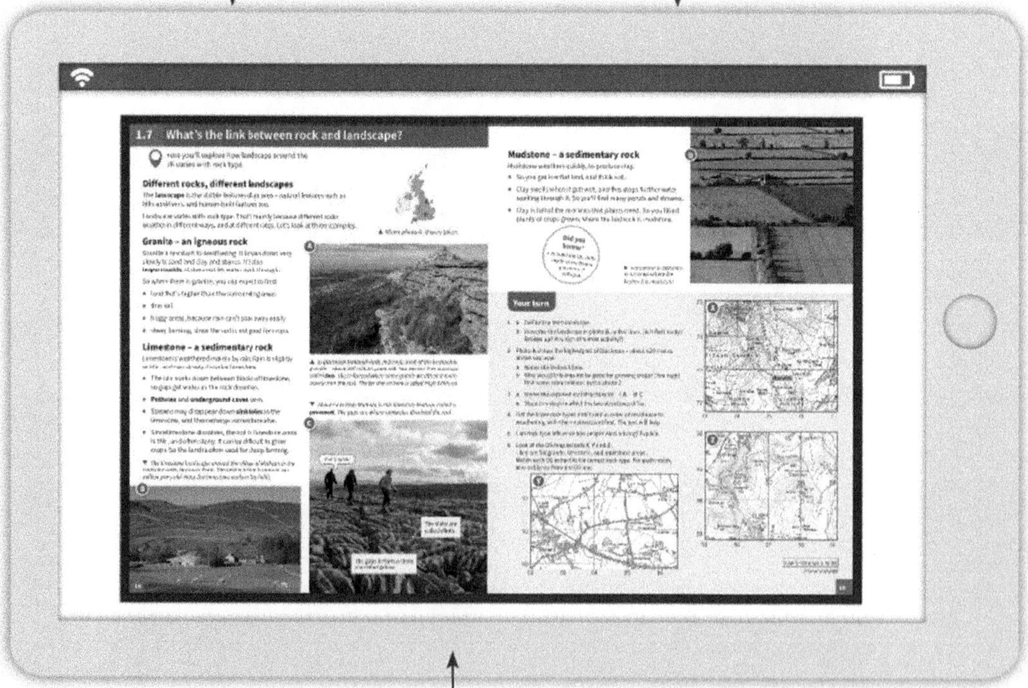


Can be accessed on a range of devices with Internet connectivity, including iPads and other tablets – so students can have access anywhere, any time.

## *Kerboodle Lessons, Resources, & Assessment*

In addition to the 'Teaching version' of the students' book, the *Kerboodle Lessons, Resources, & Assessment* includes a digital version of this teacher's handbook and a wealth of material to support and extend learning from the students' book. Kerboodle updates ensure that content remains relevant and effective.

## Lessons

There are Lesson activities for each chapter in the students' book. These activities can be completed in class or at home. They include support with key terminology, organising knowledge, and extension activities. Answer guidance is included which is only accessible to teachers.

Lesson activities can be accessed from the 'Lessons' area of Kerboodle. These can be used during the lesson as a quick check of students' understanding.

Lesson activities can also be assigned to students to complete at home.

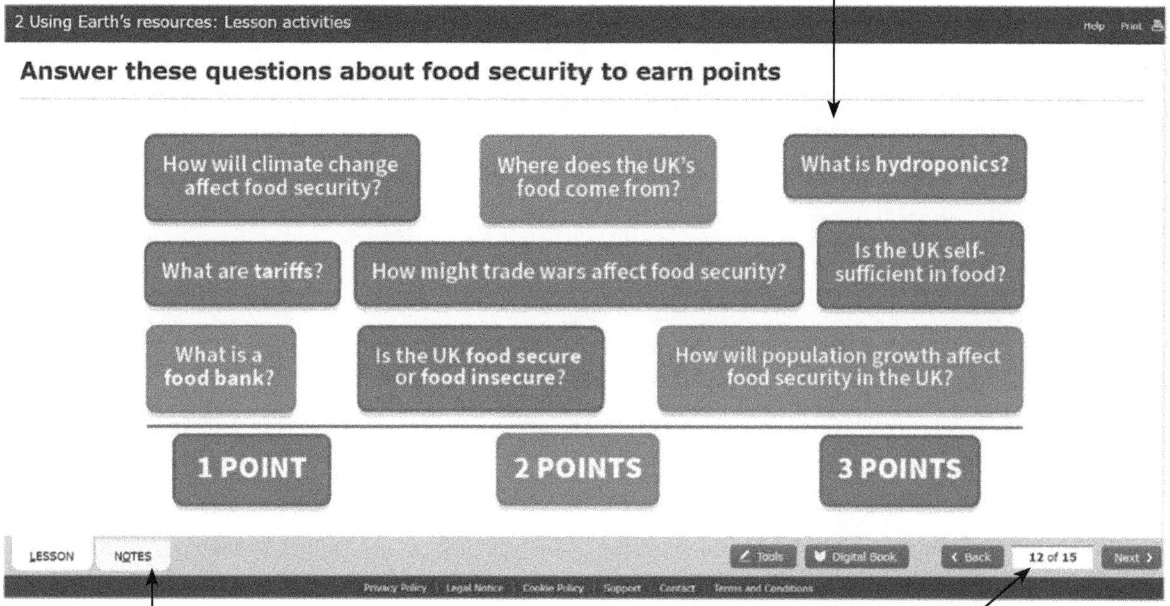


Answer guidance is available from the 'Notes' screen. This is only accessible to teachers.

There are a variety of activities for each chapter to support students with a range of learning objectives.

## Resources

All *Kerboodle* resources can be assigned to students for use at home and many of them can be adapted to suit the needs of your students. Teachers can also personalise their Kerboodle course by uploading their own resources.

- Class interactives, animations and skills-pods support students with understanding geographical concepts, skills and key terms.
- Foundation and Extension worksheets can be adapted, printed and photocopied for use in class or at home. Worksheet answers are also provided.
- Glossaries are provided for every chapter. These include key terms and definitions, helping students to write like a geographer. Glossary worksheets only include the key terms, for students to research and write their own definitions.
- Two Week Topics are provided for every chapter. These include a 3-step process to support students in selecting a topic from the students' book and planning their own project using Kerboodle resources.

# About assessment on *Kerboodle*

## Assessment

The *Kerboodle Lessons, Resources, & Assessment* includes a comprehensive range of assessment content, from no-stakes formative quizzes to summative assessment papers. All assessment content can be assigned to students to complete by a specified date.

### Checkpoint quizzes

Checkpoint quizzes are auto-marked diagnostic quizzes for each chapter. The checkpoint quiz diagnostic tool helps teachers to use the results to identify the strengths and weaknesses of the class and individuals in key topic areas. Checkpoint quiz teacher notes include guidance on how to use the checkpoint quiz resources and suggested next steps.

### Key terms quizzes

Key terms quizzes are auto-marked, low-stakes quizzes to check understanding of the key terms for each chapter.

### End-of-lesson tests

End-of-lesson tests are auto-marked quizzes to check knowledge and skills for each unit (double-page spread) in the students' book. Feedback is provided after each question to support formative learning. Each test may take no longer than 20 minutes for most students to complete.

### End-of-chapter tests

End-of-chapter tests are auto-marked quizzes to check knowledge and skills for each chapter in the students' book. Feedback is provided at the end of the test. Each test may take no longer than 30 minutes for most students to complete.

### Self-assessment checklist

Self-assessment checklists can be downloaded and photocopied. They help students to track progress on their learning goals for each chapter.

### End-of-chapter assessment papers

End-of-chapter assessment papers include KS3-appropriate exam-style questions and should take around 35 minutes to complete. They are marked using the mark scheme provided which allows for the results to be provided in several ways, to suit your requirements. Further detail about the End-of-chapter assessment papers is provided on pages 16–17.

## The markbook

The markbook provides a clear record of the assessment results for your students and helps to track progress. For auto-marked tests and quizzes, the marks are automatically entered into the Kerboodle markbook when the student submits the test. For manually-marked tests, teachers can enter the percentage score into the markbook after marking the test.

The markbook reporting tools allow you to run reports for individual students, classes, and year groups for a single assessment or across multiple assessments. You can also run diagnostic reports for the auto-marked assessments to see question-by-question performance across a whole class.

Turn over to find out more about the End-of-chapter assessment papers.

# Using the End-of-chapter assessment papers

There is an exam-style assessment paper for each chapter on *Kerboodle Lessons, Resources, & Assessment.*

Each is structured like a GCSE exam paper: short point-marked questions followed by longer level-marked questions – the final question typically carries 9 marks and provides the opportunity for extended writing. Each paper has a total of 25 marks.

They are designed to be completed in a typical lesson – you can choose how long to give your students, but they have been written with a time allowance of 30 or 35 minutes in mind.

There's a mark scheme for each paper.

You can choose how to give student outcomes:

- a mark out of 25
- a percentage
- a performance indicator – we use 'Developing', 'Secure', 'Extending', corresponding to GCSE grade expectations, but you could adapt to match any performance indicators used in your school
- a GCSE grade indicator

To convert raw marks to a performance or GCSE grade indicator, use this table:

| **Raw marks** | | | | | | | | | |
|---|---|---|---|---|---|---|---|---|---|
| 23–25 | 20–22 | 17–19 | 14–16 | 11–13 | 8–10 | 6–7 | 4–5 | 2–3 | 0–1 |
| **GCSE grades 9–U** | | | | | | | | | |
| 9 | 8 | 7 | 6 | 5 | 4 | 3 | 2 | 1 | U |
| **Performance indicator** | | | | | | | | | |
| Extending | | | Secure | | | Developing | | | |

If you choose to use the GCSE grade indicator, please appreciate that these are *suggested* grades based on grade boundaries from recent GCSE exam papers. If a student achieves a Grade 7 in one of these assessments, it is not the equivalent of a Grade 7 at GCSE. Instead it is an indicator of the grade the student could expect to get if they continue on their flight-path through KS3 and GCSE.

We would point out three things:

- the raw-mark boundaries we have used in the framework assume that students will have prepared for their assessment by doing at least some revision – if you set the paper without revision, you might take that into account when awarding GCSE grade indicators
- the raw-mark boundaries in the framework are based on but do not match precisely those for recent GCSE exam papers – this is for two reasons: firstly, our assessment papers are short, focused assessments likely to be done at the end of a chapter (topic) rather than at the end of the course; secondly, we have been careful to avoid creating papers and mark schemes that allow students to access the top grades too easily – the dangers of a false sense of security are obvious
- the exam-style papers for *geog.3* are a little more demanding than those for *geog.1* and *geog.2* in that they use higher-order command words – if you set one of them for Year 7 or Year 8 students, you might take that into account when awarding GCSE grade indicators

The reporting framework also needs to be understood as a 'best estimate'. You may choose to adapt it and apply your own grade descriptors, or to provide actionable feedback, or to simply inform students whether they are above, on, or below their target.

Each exam-style paper is a relatively short assessment, and it is common for student performance to fluctuate, often quite widely, across KS3 assessments. Therefore, in order for these grades to have value and significance as regular (termly or half-termly) screening points and/or GCSE-grade predictors, you should have a system in place. This is likely to involve using multiple assessment results to calculate one overall 'working-at' grade for each student.

Assuming that you report using GCSE grade indicators, you might:

- work out an average grade for all the assessments completed by a student (for example, since the start of the year)
- use the highest grade achieved twice by a student
- use the highest grade achieved once by a student

Of the three approaches, we have found **'highest grade achieved twice'** to be the most effective and reliable indication of a student's capabilities and progress.

If paired with 'average grade for all assessments', you would achieve a very robust indication.

These exam-style papers should reduce the possibility of wild grade fluctuations between assessments because they are all written in the same style and test both skills and knowledge-retention at a similar ratio on each paper. We are confident that grades from several will have real value as an indicator of a student's performance and progress.

## Entering student outcomes into the Kerboodle Markbook

The Kerboodle Markbook records scores and percentages to allow for quick comparison of performance.

# 1 From rock to soil

## About this chapter

### Addressing the KS3 Programme of Study

- Understand, through place-based exemplars at a variety of scales, the key processes in …
  - physical geography relating to: rocks, weathering, and soils; geological timescales
  - human geography relating to: use of natural resources
- Interpret Ordnance Survey (OS) maps in the classroom

### About the content

- We include a unit about the travels of the British Isles over the last 550 million years. It helps to explain our great variety of rock types and fossils, and our mountains.
- Early ideas of plates and plate movements are introduced. We return to them in Chapter 5.

## The big picture

These are the key ideas in this chapter.

- Rock is a mixture of natural compounds called minerals. These have chemical names, but we usually call them by their geological names.
- All rock belongs to one of three rock groups, depending on how it was formed: sedimentary, igneous, or metamorphic. Within each group are many different types of rock. For example limestone and sandstone are sedimentary rocks. Granite and basalt are igneous rocks.
- Rock can move from one group to another in a set of processes called the rock cycle.
- Within the rock cycle, rock gets broken down by weathering. Physical weathering breaks rock into bits. Chemical weathering changes the minerals.
- The end result of weathering is soil. We depend on soil for the elements that make up our bodies (which reach us through food).
- The UK has a wide range of rock types, formed in different environments. This is because the land has been on a long journey. Events on the journey also explain our mountains.
- Different rock types lead to different kinds of landscape, mainly as a result of weathering.

## The chapter outline

Use this to give students a mental roadmap for the chapter.

1 **From rock to soil** The chapter opening photo in the students' book is an important resource. See opposite for using it.

1.1 **Our rocky home** What is rock, and what do we use it for?

1.2 **What are the three rock groups?** Meet sedimentary, igneous, and metamorphic rock

1.3 **What is weathering?** The ways in which rock gets broken down, and how soil is formed

1.4 **What is the rock cycle?** How rock changes from one rock group to another

1.5 **The British Isles on their travels!** Where they've been, over the last 550 million years

1.6 **Rock and relief around the UK** The geology of the UK – and why it has all those mountains

1.7 **What's the link between rock and landscape?** How landscape varies with rock type; we use OS maps to check

1.8 **How are *you* linked to soil?** The soil profile – and why we depend on soil

1 **From rock to soil check questions** A page of questions to check knowledge, understanding and skills across the chapter, with an opportunity for extended writing

## Objectives and outcomes for this chapter

| Objectives | Unit | Outcomes |
|---|---|---|
| Most students will understand: | | Most students will be able to: |
| • that rock is made up of natural compounds called minerals | 1.1 | • say that rocks are made of minerals; explain what a mineral is, and name at least three; state that there are different kinds of rock, with different mixtures of minerals |
| • that all rock belongs to one of three groups, depending on how it was formed | 1.2 | • name the three groups (sedimentary, igneous, metamorphic) and say how the rocks in each group were formed; name at least one rock in each group |
| • that rock is broken down by weathering | 1.3 | • name the two types of weathering and give the key difference between them; describe at least two processes in physical weathering; give an example of chemical weathering; say that the weathering of rock leads to soil |
| • that rock moves from one group to another, in the rock cycle | 1.4 | • explain what the rock cycle is; describe for example how sedimentary rock becomes igneous rock, and vice versa |
| • that continents and oceans slowly move around and change shape, because they are carried on moving plates | 1.5 | • say what plates are and that they move around, describe how the land that now forms the British Isles moved over the last 550 million years |
| • that plate movements led to our mountains | 1.6 | • say that our mountains were formed when plates pushed into each other, and rock was forced upwards |
| • that Britain has rock from all three rock groups | 1.6 | • say where most metamorphic and igneous rock occurs in the UK; and that the rest is sedimentary |
| • that the features of a landscape are related to the type of rock | 1.7 | • say that different rocks lead to different landscapes, because the rocks weather in different ways and at different rates; give at least two characteristics each for landscapes in granite, limestone, and mudstone areas |
| • that we depend on soil | 1.8 | • describe and sketch the soil profile; explain how our bodies depend on elements from soil |

These tie in with the outcomes for each unit in this teacher's handbook.

## Assessment for this chapter

There are assessment materials for this chapter on *geog.3 Kerboodle Lessons, Resources, & Assessment*. They include interactive auto-marked end-of-lesson and chapter-level tests, a printable exam-style end-of-chapter paper (with a mark scheme), and a printable self-assessment checklist. See also the notes on pages 14–17 of this book.

## Getting ready for this chapter

There is a scheme of work for this chapter, plus a glossary and glossary worksheet, on *geog.3 Kerboodle Lessons, Resources, & Assessment*.

## About the chapter-opening photo

Ask students:

How do you think that mountain formed?

Will it look the same 20 million years from now?

The image shows Mount Suilven on the north-west coast of Scotland.

The mountain sits in a landscape of Lewisian gneiss (pronounced nice), a rock about 3 billion years old. Mount Suilven itself is made of ancient sandstone, deposited around 800 million years ago in a layer up to 1 km thick or more. The sandstone was then sculpted by glaciers, leaving the mountain sticking up.

# Our rocky home



## About this unit

This unit explains what rock is, what it is made of, and what we use it for.

## Key points

- Rock forms the Earth's crust, and was first formed over 4.5 billion years ago.
- Rock consists of a mixture of minerals (natural compounds) which exist in rock as crystals.
- The most common rock-building minerals are feldspars and quartz.
- Rocks are used for many things, including building, to obtain metals, and to obtain gemstones.

## Key vocabulary

rock, minerals, crystals, granite, quartz, feldspars, sandstone, limestone, calcite, chalk, rock-building minerals, crust, concrete, cement, aggregate, metal ores, gemstones

## Skills practised in 'Your turn'

- Literacy skills: q2, q6, define; q8, write a dramatic article (10+ lines) for a news website
- Thinking skills: q1, q3, q6, comprehension; q3, q5, q7, give examples; q4, unjumble a word; q8, think about consequences

## Unit outcomes

By the end of this unit most students should be able to:

- explain the terms given in 'Key vocabulary' above
- say when Earth's rock first formed
- explain what minerals are, and name the most common rock-building minerals
- give three ways that the building industry uses rock
- explain what a metal ore is

## Ideas for a starter

1. With books closed, display a set of images that all have one thing in common – they are rock-related. For example, abseiling / rock climbing, stone being quarried, a sculptor at work, a diamond cutter at work, Everest, tors, etc. Ask: What is the link?
2. With books closed, write anagrams of rock names on the board for students to decipher. For example: tearing (granite); tie melons (limestone); teals (slate).

## Ideas for plenaries

1. Show images of different kinds and colours of rock to illustrate the great variety. You could include diamonds and other precious stones. An Internet search for *close up of minerals in rock* could be a good starting place.
2. Ask: Why does granite have different colours in it?
3. Ask: Why bother learning about rocks in geography?
4. Ask: Do you think rock could melt?

## Further class and homework opportunities

Suggestions 1–3 on page 37 of this book

*geog.3 workbook,* page 4

**kerboodle**

### Resources

1.1 You decide! – Students rank the importance of rock uses.

1.1 Sort it out! –Students complete the features of rocks.

1.1 Foundation worksheet – Students write a statement identifying uses for rocks and match rock types to their descriptions.

1.1 Extension worksheet – Students list rock types, their uses, and the benefits and problems of using rock.

### Assessment

1.1 End-of-lesson test – An assignable, auto-marked formative assessment.

## Answers to 'Your turn'

**1** Over 4.5 billion years ago

**2** Minerals are natural compounds found in rock. They are solid, and usually exist as crystals.

**3** **a** Quartz

**b** Sandstone and limestone (and granite)

**4** Chalk

**5** Any three from: to construct buildings made from blocks cut from rock; rock and sand is used to make concrete; limestone is used to make cement; crushed rock (or aggregate) is used in road building; slate is used as roof covering; rock is sometimes used for worktops and tiles.

**6** **a** The rocks that metals come from.

**b** Bauxite

**7** Answers will vary but students are likely to suggest: the buildings they live in and use are built from rock, concrete, cement, and metal; surfaces such as roads, footpaths, and the school playground are made from rock; metal is used to make cars, planes, trains, mobile phones, and household objects such as fridges and microwaves; other electronics such as televisions and laptops have metal components. Accept labelled sketches or spider diagrams.

**8** There's plenty of scope for students' imagination here. You'd expect structures made of rock to get eaten up (some buildings, pavements, walls, gravestones, statues, slate roofs). The fungus would attack exposed rock on hillsides and mountains and work its way down till it reached Earth's metal core. (You could direct students to the first diagram on page 88 of their books.) Eventually, even if the fungus did not eat concrete, all built structures would topple as the rock below disappeared. And if Earth got lighter, its place in the solar system would be affected too.

# What are the three rock groups?



## About this unit

This unit introduces the three rock groups – sedimentary, igneous, and metamorphic – and shows how the rocks in each group were formed. It also gives their relative percentage in Earth's crust.

## Key points

- All rock belongs to one of three groups of rock: sedimentary, igneous, and metamorphic.
- Sedimentary rock forms when a thick heavy layer of sediment gets compacted, and its particles get cemented together.
- Igneous rock forms when any kind of rock melts, then cools and hardens again. It may cool and harden underground, or on Earth's surface.
- Metamorphic rock forms when any rock is changed by heat and pressure, without melting. Usually the structures of the minerals in the rock undergo change.

## Key vocabulary

sedimentary rock, metamorphic rock, igneous rock, mudstone, sandstone, limestone, magma, lava, granite, basalt, slate, tuff

## Skills practised in 'Your turn'

- Geographical skills: q1, draw a flow chart
- Thinking skills: q1, suggest a name; q2, q4, explain / give reasons; q3, explain a process; q3, q5, comprehension

## Unit outcomes

By the end of this unit most students should be able to:

- explain the terms given in 'Key vocabulary' above
- describe how the rocks in each of the three groups are formed, and name at least one rock in each group
- say what is similar, and what is different, about granite and basalt
- explain the connection between mudstone and slate
- say that over half of the rock (65%) in Earth's crust is igneous rock; only about 8% is sedimentary rock, and it forms a layer over most of Earth's surface

## Ideas for a starter

1 With books closed, pass around some very different samples of rock. For example granite, basalt, limestone, pumice, marble, sandstone. Ask for suggestions about how they were formed. You could write notes on the board. Say you will come back to these later.

2 With books closed, write IGNEOUS ROCK, SEDIMENTARY ROCK, and METAMORPHIC ROCK on the board in big letters. Explain that they are the names of the three big groups of rock. All the rocks in each group were formed in the same way. Can anyone guess how the rocks in each group were formed?

## Ideas for plenaries

1. What makes slate good as a roofing material? (Thin, flat sheets which can be cut easily, hard, does not let water through.)
2. What's the difference between a rock and a fossil?
3. If you chose starter 1, revisit the rock samples now.
4. Do you think rock is still being formed in each of those groups? Why do you think that?

## Further class and homework opportunities

Suggestions 4–7 on page 37 of this book

*geog.3 workbook*, page 5

**kerboodle**

### Resources

1.2 Skills workshop – Students explain the formation of sedimentary rock.

1.2 Quiz time! – Students complete a quick quiz about igneous rock.

1.2 Foundation worksheet – Students answer questions on the three rock groups and describe how mudstone changes over time using words and pictures.

1.2 Extension worksheet – Students define the rock groups, complete an odd-one-out activity, and outline the process of sea shells becoming marble.

### Assessment

1.2 End-of-lesson test – An assignable, auto-marked formative assessment.

## Answers to 'Your turn'

**1 a** Students should use the text and the three images / diagrams from the top of page 8 to create a flow chart, with arrows dividing each separate process.

**b** The rock is in fact called siltstone.

**2 a** Limestone forms from shells and other hard remains of organisms that pile up on the ocean floor. Chalk forms in the same way as limestone, but from very tiny organisms.

**b** Yes, they are still forming because shells and organisms continue to form layers on the ocean floor. As the layers get thicker and heavier, they compress the material below. To form new rock, the process may take millions of years.

**3 a** Marble is a metamorphic rock.

**b** If marble heats up under the ground (due to hot magma from the mantle), it could melt. When it cools and hardens again it would become igneous rock.

**4 a** As it is pushed down, the sedimentary rock will have changed into other forms of rock: metamorphic or igneous.

**b** The rock they were trapped in will have melted before it hardened to form igneous rock. So you'd expect the fossils to have been destroyed.

**5** Sedimentary rock.

# What is weathering?



## About this unit

This unit presents the two types of weathering: physical and chemical. It describes the processes. Then it looks at the end result – soil – and factors that affect the rate of weathering.

## Key points

- Weathering is the blanket term for the way rocks are broken down by things in their environment: heat, cold, rain, gases in the air, and so on.
- There are two types of weathering – physical and chemical. (Some people add a third – biological – but we include this in 'physical'.)
- In physical weathering rock is broken into smaller bits, but the minerals in it don't change.
- In chemical weathering, minerals in the rock undergo chemical reactions, and change. This helps to break the rock down.
- In the end, rock breaks down to soil.
- How quickly rock weathers depends on the climate, and the rock.

## Key vocabulary

physical weathering, chemical weathering, freeze-thaw weathering, exfoliation, feldspars, clay, quartz, soil

## Skills practised in 'Your turn'

- Geographical skills: q2, draw diagrams
- Literacy skills: q1, q5, word association
- Thinking skills: q3, identify key difference; q4, q5 describe a process; q5, q6 explain a process

## Unit outcomes

By the end of this unit most students should be able to:

- explain the terms given in 'Key vocabulary' above
- name the two types of weathering, and state the key difference between them
- describe four processes in physical weathering, and two in chemical weathering
- say that the weathering of rock leads to soil
- give the two key factors that affect the rate of weathering (climate and rock type)

## Ideas for a starter

1. Bring in a jar of soil. Ask: What is this stuff? Where do you think it came from? Do you think it has been here since Earth began? Say they'll find out more during the lesson.
2. With books closed, display images of weathered rock. For example, stone statues and gargoyles which have lost their features, and illegible writing on gravestones. What has caused these changes? Elicit that it could be the effect of air, rain, temperature, and so on.

## Ideas for plenaries

1. Look at the photo at the top of page 10 of the students' books. What kinds of weathering may be going on in this place?
2. Give me five! Summarise today's lesson in five points, five words in each. (Students could work in pairs for this.)
3. Students create a spider diagram on the board, under instruction from the class, showing how rocks are weathered.

## Further class and homework opportunities

Suggestions 8 and 9 on page 37 of this book

*geog.3 workbook*, page 6

**kerboodle**

### Resources

1.3 Sort it out! – Students categorise the characteristics of physical and chemical weathering.

1.3 Quiz time! – Students identify the types of weathering from photos.

1.3 Foundation worksheet – Students complete a paragraph about weathering and identify the conditions and key features of physical, chemical and biological weathering.

1.3 Extension worksheet – Students write a fictional account detailing a rock's journey through various weathering processes until it eventually becomes soil.

### Assessment

1.3 End-of-lesson test – An assignable, auto-marked formative assessment.

## Answers to 'Your turn'

**1** The name was chosen because it is mostly due to elements of weather: temperature and rainfall.

**2** Sample drawings are shown on the right.

**3** *Physical weathering* breaks rock down into smaller bits – but the substances in the bits do not change. *Chemical weathering* changes the substances into different substances, through chemical reaction.

**4** By the reaction of water with feldspars in rock. (The reaction with water is called *hydrolysis*.) Remember, feldspars are very common in rock, so clay is a very common product.

**5** **a** Weathering by living things is often called biological weathering. This is because 'biology' relates to living organisms.

**b** Burrowing animals like rabbits can widen cracks in rock by forcing their way through.

**c** Lichen grows into cracks in rocks and widens them; it also makes acids which eat into rock.

**6** Limestone rock weathers more slowly in dry areas because rain is slightly acidic, and it can dissolve rock. Rain is acidic because it dissolves carbon dioxide from air – which then reacts with the calcium carbonate in limestone which in turn, dissolves the rock.

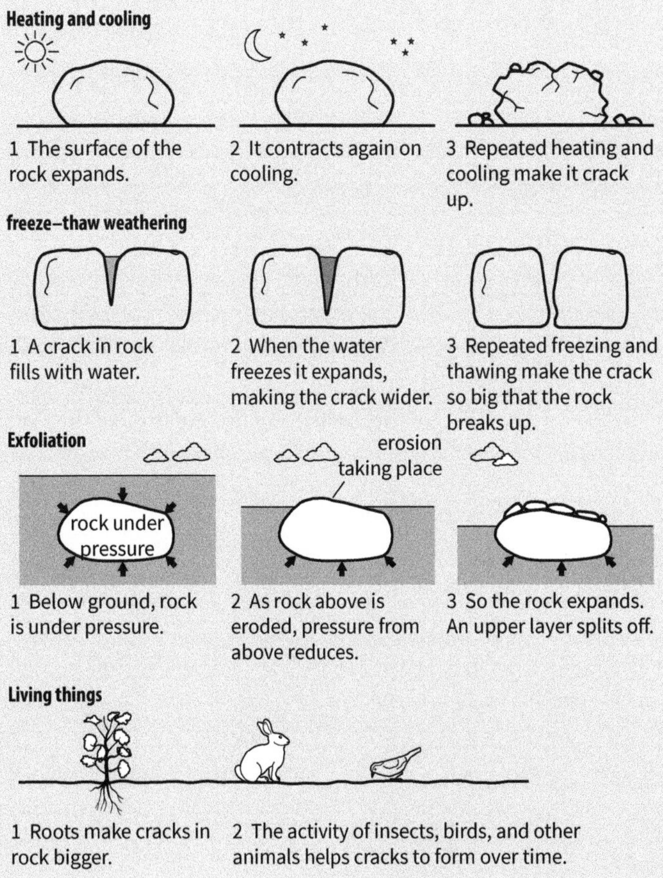

# What is the rock cycle?



## About this unit

This unit introduces the rock cycle, in which rock changes from one rock group to another. It looks at what drives the rock cycle, and how it helps us.

## Key points

- The rock cycle is the set of processes which result in rock changing from one rock group to another (sedimentary, igneous, metamorphic).
- The cycle continues non-stop.
- The cycle is driven by plate movements.
- The rock cycle helps us by forming rocks which can be used as building materials, metal ores, and gemstones.

**Note:** the work on plates and plate movements is developed in Chapter 5. There are many opportunities to make links between Chapters 1 and 5.

## Key vocabulary

rock cycle, weathering, erosion, transportation, deposition, sedimentary / metamorphic / igneous rocks, uplift, magma, lava, plates, plate movements

## Skills practised in 'Your turn'

- Geographical skills: q8, interpret a photograph
- Literacy skills: q4, define
- Thinking skills: q1, q5, explain a process; q2, match pairs of rocks by type; q3, q6, suggest a process; q7, write a summary

## Unit outcomes

By the end of this unit most students should be able to:

- explain the terms given in 'Key vocabulary' above
- outline what happens during the rock cycle and draw a rough sketch
- describe how plate movements drive the rock cycle
- state two ways that the rock cycle helps us

## Ideas for a starter

1. With books closed, draw some boulders on the board. Ask: What will happen when these get weathered? (They'll be broken down to soil.) So will we run out of rock one day? Talk around this question, but without mentioning the rock cycle.
2. With books closed, pass around (or draw on the board) a piece of sedimentary rock, eg sandstone or limestone, and a piece of igneous rock, eg basalt. Ask how each was formed. Then say that in nature, sedimentary rock may change into metamorphic and igneous rock, and vice versa. How could this happen? You could note ideas on the board.

## Ideas for plenaries

1. Rock is being broken down non-stop, by weathering. Then the material is eroded by rivers (and wind, rain, and ice). So why is Earth not getting smaller?
2. How long do you think it takes for an atom from rock to go all the way from 1 to 6 on the rock cycle diagram, and back to 1? (Many millions of years.)
3. Stuck in the box! Place a large cardboard (or imaginary) box at the front of the class. Select a student to stand in the box. They can leave the box only after answering correctly a question about today's lesson. (It could be a question about vocabulary or concepts.)

## Further class and homework opportunities

Suggestions 10–12 on page 37 of this book

*geog.3 workbook*, page 7

**kerboodle**

### Resources

1.4 Sort it out! – Students match the sedimentary and metamorphic rocks.

1.4 Quiz time! – Students complete an explanation of the rock cycle.

1.4 Foundation worksheet – Students draw and label the rock cycle and apply this knowledge to how sea shells turn to marble.

1.4 Extension worksheet – Students draw and label the rock cycle, and go on to explain how plate movements drive the cycle.

### Assessment

1.4 End-of-lesson test – An assignable, auto-marked formative assessment.

## Answers to 'Your turn'

1. Over millions of years, all rock on Earth's surface will be broken down by weathering processes into stones and sand. These are small enough to be carried away by rivers, glaciers or wind. Eventually the weathered rock will be deposited in the ocean and will form a layer of sediment. When compacted under the weight of the ocean and beneath other layers of sediment, it will become sedimentary rock.
2. Limestone and marble; sandstone and quartzite; mudstone and slate.
3. It is made from a sediment of shells and coral piling up on the ocean floor.
4. Plates are slabs which make up the hard, outer part of Earth's surface.
5. At plate boundaries, there are certain conditions that can cause rock to melt. Diagram B shows a destructive (converging) plate boundary, where one plate is submerged below another and as it sinks underground, the heat and pressure change it to metamorphic rock.
6. Uplift can raise limestone up from the ocean floor, so it can become land.
7. Students should summarise the content from diagram A and the text below it. Good answers will include the terms in bold.
8. The caption 'All thanks to the rock cycle' relates to the buildings and the vehicles in the image. Buildings are made of rock, cement, concrete, metal, glass; and these are all products of the rock cycle. Cars and buses are made from metals, which are also extracted from rocks, as metal ores. However, some students might comment that there are some other materials, which are not part of the rock cycle – such as plastics, which are a compound made from various natural materials such as cellulose, salt, and oil.

# 1.5 The British Isles on their travels!

Help at a glance

## About this unit

This unit describes how the land which forms the British Isles has travelled the globe over millions of years to its present position – thanks to plate movements.

## Key points

- Earth's continents and oceans are carried on plates.
- As plates move, they sometimes join other plates, and sometimes break apart.
- So the continents and oceans may change their shapes, as well as their positions.
- For example, the land that now forms the British Isles was once in the southern hemisphere, and on two separate continents. It came together when plates pushed into each other – and was then carried north to its present position.
- As it travelled, the land spent millions of years in different environments, that had different species. That's why the UK has fossils from faraway places.

## Key vocabulary

plates, Pangaea, fossil, geological timescale, Precambrian eon, era, period

## Skills practised in 'Your turn'

- Geographical skills: q2, q3, interpret maps and lines of latitude; q4–7, interpret geological timescale
- Numeracy skills: q4, q6, use the geological timescale; q8, calculate foot size in cm
- Thinking skills: q1, q7, explain; q2, give differences; q3, suggest climatic conditions

## Unit outcomes

By the end of this unit most students should be able to:

- explain the terms given in 'Key vocabulary' above
- say that continents, and oceans, are carried on plates; as the plates move, and join, and break up, the positions and shapes of the continents and oceans change too
- describe the northward journey of the land that is now the British Isles, from 550 million years ago (mya)
- explain why dinosaur fossils found in the UK are all more than 66 million years old

## Ideas for a starter

**1** With books closed, display an image of any fossil found in the UK from around 350 mya, when the land was near the Equator. For example a fossilised rainforest tree from Forest Grove in Glasgow, or fossils from ancient coral reefs found in south Wales. These species lived in warm tropical places. So why were their fossils found in the UK?

**2** With books closed, say: 350 million years ago, the land you are sitting on was close to the Equator. Can you explain this mystery?

## Ideas for plenaries

1 Ask questions about the geological timescale. For example: What do *mya* and *bya* mean? Where does the scale begin, at the top or bottom? When did Earth form? In which period did humans first appear? In which periods were there dinosaurs? When did coal begin to form? (In the Carboniferous Period.) About when did the Jurassic Period begin, and how long did it last?

2 The plate which carries the UK and the rest of Europe is still travelling (eastwards). Today scientists can measure its speed and direction. How do you think they do that?

## Further class and homework opportunities

Suggestions 13–15 on page 37 of this book

*geog.3 workbook*, page 8

**kerboodle**

### Resources

1.5 Where in the world? – Students complete a description of continental plate movements.

1.5 Word scramble – Students complete a crossword of the geological time periods.

1.5 Foundation worksheet – Students answer true or false statements about the British Isles and put the geological time periods in order, matching them to key events.

1.5 Extension worksheet – Students fill in gaps on a geological timescale and explain why the land that's now the UK has been under ice and warm tropical seas.

### Assessment

1.5 End-of-lesson test – An assignable, auto-marked formative assessment.

## Answers to 'Your turn'

1 Expect students to state that those continents no longer exist because the plates they were carried on have moved around, have joined together, and broken apart, over many millions of years (as they are still doing today).

2 Any two from: the British Isles was part of the land mass of Europe but now it has broken away; India was once an island but today it is part of Asia; South America was once separate to North America but now they are joined by land; Africa and Asia were separate, but today they are connected by land in the Middle East; North America was at a high latitude but now the land mass is further south.

3 Students should describe the journey of the British Isles from the southern hemisphere to the northern hemisphere. Good answers will refer to lines of latitude and names of land masses / continents. Five hundred and fifty mya the climate would have been colder than today as the British Isles was located close to the Antarctic Circle. Between 395 and 300 mya would have been warmest as the British Isles crossed the Equator on its journey north.

4 Life began in the sea.

5 Plants appeared on land before animals.

6 Birds appeared in the Jurassic period which was between 200 and 145 mya.

7 Dinosaurs became extinct at the end of the Cretaceous period – 66 mya.

8 **a** Triassic

**b** Pangaea

**c** A student who wears a size 6.5 shoe would have the same length footprint (23 cm). However, we have five toes compared to the four belonging to this reptile; and the toes and nails are shorter on human feet.

# Rock and relief around the UK

Help at a glance

## About this unit

In this unit students will find out why the UK has mountainous regions, and different types of rocks.

## Key points

- Two key factors explain why the UK has mountainous and hilly regions – plate movements and rock type.
- The mountains formed when plates collided, causing land to fold upwards creating mountains.
- The bedrock of the UK consists of igneous, metamorphic and sedimentary rocks – but most of it is sedimentary. (Sediment was laid down at different times and in different environments, as the British Isles travelled.)
- Different rock types weather at different rates; higher land (eg in Scotland) is a sign that the rock is more resistant to weathering; low land is a sign that rock weathers more easily.

## Key vocabulary

relief, plate movements, rock type, bedrock, weathering

## Skills practised in 'Your turn'

- Geographical skills: q2, compare a relief map with a geology map; q3, q5, q6, interpret a photo; q4, interpret a geology map
- Numeracy skills: q6, use timescale
- Literacy skills: q4, define
- Thinking skills: q1, q3, explain

## Unit outcomes

By the end of this unit most students should be able to:

- explain the terms given in 'Key vocabulary' above
- say which parts of the UK are most mountainous, and how and why the mountains formed
- name the most common type of bedrock in the UK
- name three types of sedimentary rock found in the UK
- describe the link between rock type and relief

## Ideas for a starter

1. With books closed, draw a rough outline map of the UK on the board. Ask students to come up and mark in where they think the most mountainous parts are. Which can they name?
2. Recap: Ask for a volunteer to give a 30-second review of what the last lesson covered. Ask others to fill in the gaps. Use prompts to remind students of the key ideas.

## Ideas for plenaries

1. Ask questions to encourage students to compare the relief map and the geological map. For example: Where is most of Britain's highest land, and what types of rock is it made of? Some of the metamorphic rock in Scotland is up to 2.5 billion years old. So what can you say about the rate of weathering of metamorphic rock? Where is the bedrock where you live? (N.B. a map at a larger scale might give a different answer.)
2. The sedimentary rock in the UK is of different types and ages. Who can explain why?

## Further class and homework opportunities

Suggestions 16–18 on page 37 of this book

*geog.3 workbook*, page 9

**kerboodle**

### Resources

1.6 Where in the world? – Students match the upland areas of the UK with their rock types.

1.6 Sort it out! – Students explain how rock type affects relief.

1.6 Foundation worksheet – Students complete the gaps in a paragraph explaining why there are different rock types and relief in the UK.

1.6 Extension worksheet – Students write a paragraph using geographical terminology to describe rock types and relief in the UK.

### Assessment

1.6 End-of-lesson test – An assignable, auto-marked formative assessment.

## Answers to 'Your turn'

**1** Because rock got folded upwards as land masses were forced together, when plates collided. (Look back at globes **2** and **3** on page 14 of the students' book.)

**2** **a** Grampian Mountains

**b** Metamorphic rock

**c** Higher land is a sign that rock is resistant to weathering, and there are mountains here. (But students should also remember that mountains gain height when the land is squashed as plates collide.)

**3** **a** Students should notice that the rock has buckled and folded.

**b** This is metamorphic rock which is 3 billion years old and is the oldest in the UK. Its age and status indicate that it is very resistant. Less resistant rock would not have lasted this long, especially at the coast, where it is under attack from the sea.

**4** **a** The solid rock underlying the soil.

**b** Sedimentary rock is more common, followed by metamorphic, then igneous.

**5** Mudstone weathers more quickly and easily than sandstone. There is evidence from the photo (and the text below it) that the higher ground in the background is capped by sandstone, which suggests it has been more resistant to weathering.

**6** **a** The top layer has the youngest rock because these cliffs are made of sedimentary rock, which forms in layers on the ocean floor. The top layers contain the most recent deposits of sea creatures, formed under deep oceans.

**b** The oldest fossils would be in the bottom layer.

# 1.7 What's the link between rock and landscape?

Help at a glance

## About this unit

This unit explores the link between rock type and landscape.

## Key points

- Landscape varies with rock type. This is mainly because different types of rock weather in different ways, and at different rates.
- Granite is resistant to weathering. It breaks down very slowly into sand, clay, and stones.
- Limestone is dissolved by rain – because rain is slightly acidic, and acids react with carbonates. (Limestone is mainly calcium carbonate.)
- Mudstone weathers quickly, to produce clay.

## Key vocabulary

landscape, granite, impermeable, limestone, mudstone, pavement, pothole, cave, sinkhole, tor

## Skills practised in 'Your turn'

- Geographical skills: q1, q3, describe from a photograph; q2, q3, use evidence from a photograph; q6, interpret OS maps and match them to rock types
- Literacy skills: q1, define
- Thinking skills: q2, q5, explain; q3, compare; q4, put in order; q6, match and give evidence

## Unit outcomes

By the end of this unit most students should be able to:

- explain the terms given in 'Key vocabulary' above
- say that different kinds of rock lead to different landscapes
- give at least two landscape characteristics for each of these rock types: granite, limestone, mudstone
- match OS map extracts to the correct rock type

## Ideas for a starter

1. Recap, with books closed. Make some true / false statements about the work in the last lesson, for students to vote on. See who can correct the false statements. (For example: Most of the rock covering the UK is igneous rock.)
2. With books closed, display a set of images of very different inland rural landscapes in the UK. Ask students to describe what they see. (A student could write terms on the board.) Then say: Those places are all in the UK. Why are they so different?
3. Invite students to come up and draw landscapes on the board, from memory (for places they have visited), or failing that, from imagination. This will prompt a discussion of what the term *landscape* means.

## Ideas for plenaries

1. Before question 6, quiz students about the OS maps, as revision. For example: Which map has the flattest land? How can you tell? Give me a 4-figure grid reference for a square with steep land, on any of the maps. Which map has most cliffs / crags? Which map has the lowest land? How can you tell? Is it possible to say which map has the highest land, overall? Explain. Which map has most farms? Why don't the others have farms? The tor in photo A is called High Willhays. It's shown on one of the maps.

Which one? One of the maps shows marshy boggy land. Give me the 4-figure grid reference.

Note that there is an OS map key on page 138 of the students' book.

2 Using the OS maps on page 19 and the photos, create a spider map for each rock type, showing the landscape features and characteristics associated with it.

## Further class and homework opportunities

Suggestion 19 on page 37 of this book

*geog.3 workbook*, page 10

**kerboodle**

### Resources

1.7 Skills workshop – Students match the landforms to the OS maps.

1.7 Sort it out! – Students complete the description of granite landscapes.

1.7 Foundation worksheet – Students describe granite, limestone and mudstone landscapes, and identify the landscape in two photos.

1.7 Extension worksheet – Students describe and explain the features of different landscapes.

### Assessment

1.7 End-of-lesson test – An assignable, auto-marked formative assessment.

## Answers to 'Your turn'

**1 a** The landscape is the visible features of an area – natural features such as hills and rivers, and human-built features too.

**b** The landscape is undulating with gentle, grassy slopes that have been grazed short by sheep. The stone walls, the farm buildings and the livestock are all evidence of human activity. There is a steep, craggy area in the top left of the photo which looks as though it has been left wild.

**2 a** Granite

**b** Granite is impermeable so the ground above it is often boggy. It is also associated with thin soils which are not good for growing crops. The photo shows rocky outcrops, which would also make farming difficult as these would obstruct farm machinery.

**3 a i** Tor **ii** Pavement

**b** The granite has a rough surface, but the limestone is smooth and flat. There are cracks in the granite which run horizontal to the ground, whereas the limestone has deeper, wider, vertical gaps between each block.

**4** Granite, limestone, mudstone

**5** Yes. Certain rock types are better for farming – mudstone weathers to produce clay which provides nutrients for plants. Places which are rocky or mountainous are harder for people to use for economic purposes as they may be remote or hard to build on.

**6** *Evidence that X shows a limestone area:*

- disappearing streams, eg in squares 7572, 7573, 7673, 7674 (Gaping Gill in 7572 is a famous spot where a river plunges down into a cavern)
- potholes, eg in squares 7372, 7472, 7572, 7672
- caves in square 7673.

But note that the steep slopes (contour lines close together) indicate that limestone is not that easily/rapidly weathered, overall. The main effect is where water runs down the joints.

*Evidence that Y shows a mudstone area:*

- the land is flat (contour lines far apart), suggesting that this rock weathers readily
- many farms, suggesting thick soil (containing clay).

*Evidence that Z shows a granite area:*

- steep slopes (granite does not weather easily);
- many tors are marked; one is High Willhays, also shown in photo A.
- the reservoir indicates that the rock is impermeable
- there is boggy land in square 5889, suggesting that the water can't drain away.

# How are *you* linked to soil?



## About this unit

This unit looks more closely at soil, and why it is so important to us.

## Key points

- Soil is a mixture of clay, sand, and rotting vegetation.
- Soil is rich in living things, and particularly bacteria.
- If you dig down through soil you will find different layers. These make up the soil profile.
- At the top is a thin layer of humus or rotting vegetation. At the bottom is the bedrock, which weathers over time to give soil.
- Plants need elements from the minerals in soil. We call them nutrients.
- As crops remove these elements, we can add fertilisers to the soil to replace them.
- We take in these elements from the soil via the food we eat. We need them to survive.

## Key vocabulary

soil profile, humus, nutrients, topsoil, subsoil, bedrock, weathering, biodiverse, fertiliser

## Skills practised in 'Your turn'

- Literacy skills: q1, q2, q5, define
- Thinking skills: q1, q2, q4, explain; q1, suggest a reason; q3, comprehension; q6, predict

## Unit outcomes

By the end of this unit most students should be able to:

- explain the terms given in 'Key vocabulary' above
- say that soil has different layers, and name and describe them
- explain the link between soil and rock
- explain why soil is important to us, and why we should look after it

## Ideas for a starter

1. With books closed, hold up a jar of soil. Ask lots of quick questions. For example: What is it? Was it here since Earth began? How was it formed? What's in it? Why is it really important to us? Do you think there's any life in it? (Over 1 billion bacteria per teaspoon!)
2. With books closed, show a set of images of plants growing in the UK: crops, fruit trees, flowers, grass in pasture land, trees, houseplants. They are all standing in the same kind of stuff. They can't live without it. What is it? Where did it come from?

## Ideas for plenaries

1. So ... why are earthworms our best friends?
2. With books closed, ask a student, or students, to draw a pictorial flowchart on the board, beginning with rock and ending with a human, to show how elements in minerals get passed along.
3. Give a slip of paper to each student. Half of the slips have definitions or descriptions. For example 'thin top layer of soil'. The other half have the matching terms. For example 'humus'. Randomly call out a student's name. The student reads out what's on their slip. The student with the matching slip responds. You could cover material from earlier lessons too. Make sure each pair is unique, to avoid more than one student responding.

## Further class and homework opportunities

Suggestions 20–23 on page 37 of this book

*geog.3 workbook*, page 11

**kerboodle**

### Resources

1.8 Word scramble – Students complete a crossword about rock and soil.

1.8 Sort it out! – Students match the soil layers to their features.

1.8 Foundation worksheet – Students write questions on the soil profile and explain, using words and pictures, how soil has been important in the production of their breakfast.

1.8 Extension worksheet – Students formulate their own questions on the soil profile and consider the impact soil has on their day.

### Assessment

1.8 End-of-lesson test – An assignable, auto-marked formative assessment.

## Answers to 'Your turn'

1. **a** It's a mixture of clay, sand and rotting vegetation.
   **b** The clay and sand in the soil form when rock is broken down by chemical weathering.
   **c** Soils which are on mudstone are more fertile (able to produce healthy crops), because mudstone weathers quickly to produce clay which contains nutrients. Rock with higher concentrations of nutrients such as potassium and phosphorus will also weather to become more fertile.
2. Humus is a thin layer of rotting vegetation. As it rots away ('eaten' by bacteria), the nutrients in it are returned to the soil, so they can be used by other plants.
3. The topsoil
4. Soil is essential for humans because it provide the nutrients for food to grow. We need food to build up our bones, teeth, flesh, hair, etc. If we eat animal products, the nutrients come from the food eaten by the animals (grass, corn, soy, etc.), which are also reliant on soil.
5. **a** Each crop takes nutrients from the topsoil. So eventually it may not have enough nutrients to support further crops.
   **b** Fertilisers are substances added to soil to make it more fertile; they can be chemicals made in factories, or materials such as animal manure and rotting vegetation.
   **c** Animal manure
6. Rock will continue to weather, so in theory we won't run out of soil. But as the population grows, we need more soil for growing food. At the same time, as students will see in Unit 2.6, we are not looking after it. In some places, soil is being degraded far faster than new soil can form. So these places are already short of fertile soil.

# From rock to soil



## Skills practised in check questions

- Geographical skills: q1, interpret a photo; q2, draw and label a diagram
- Literacy skills: q1, q2, q3, define; q6, extended writing
- Thinking skills: q1, q2, q3, q5, explain; q1, q2, comprehension; q1, q2, identify a process; q2, q4, identify rock type; q3, q4, make a link; q4, q5, suggest a reason; q5, apply own knowledge

## Answers for check questions

**1 a** Limestone forms from a sediment of shells and coral which piles up on the ocean floor. Over years the layers build up and the weight causes the particles to squeeze together. Substances dissolved in the water act as cement and the result is new rock.

**b** Chalk

**c** One from: as a building material; to make concrete; for drystone walls.

**d** Weathering is when rock is broken down by the action of things in its environment: by heat and cold, rain, gases from the air, and even by plants and animals.

**e** When rain falls on limestone this is chemical weathering, as the rock undergoes chemical change when the calcium carbonate in the rock is dissolved by the slight acidity of the rain.

**f** Two from: pavement, potholes, underground caves, sinkholes.

**g** There is also biological weathering taking place in photo A because there are trees and shrubs growing through cracks in the rock. As the plants grow, these roots will widen the cracks.

**2 a** Students should make a copy of diagram B.

**b** See completed diagram below for rock labels.

**c** See completed diagram below for processes.

**The rock cycle**

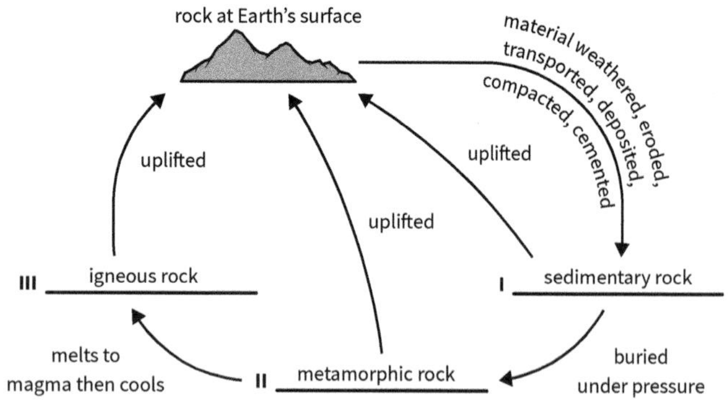


**d** Uplift occurs at plate boundaries due to pressure. When two plates push together, some rock gets buried but other rock is uplifted. This means it is pushed upwards. The ocean floor can become land as limestone is pushed above sea level. This land may eventually form mountains.

**e** Metamorphic rock

**f i** Fossils are the remains or traces of ancient life preserved by natural processes.

**ii** Fossils are millions of years old so they may depict life that existed when the continents were in different places. Some land which was at the Equator has moved north.

**3 a** Clay forms due to chemical weathering. Water reacts with feldspars (minerals) in rock to produce the fine particles that make up clay.

**b** An ore is a rock which has metals occurring naturally within it. Metals can be extracted from these rocks.

**c i** A mineral is a natural compound in rock. Rock is a mixture of different minerals.

**ii** Quartz (with some feldspars)

**d** Wheat is a crop that is farmed for food. It can be found in many foods such as bread, biscuits and cereals. It is linked to rock because it needs soil to grow. Soil includes humus from plants, and also minerals from broken down rock.

**e** Earthworms improve soil in two ways: they digest leaves and clay, so their waste is rich in nutrients; and they also create tunnels which allow air and water to circulate. Good soil means more nutritious food.

**4 a** Over time and due to the rock cycle, most rock melts under heat and pressure and changes to become a new form of rock. Therefore, rock which is billions of years old is very rare.

**b i** Rock which has just emerged from a volcano could be one day old. Once it reaches Earth's surface it cools quickly to become new rock.

**ii** Igneous rock

**5 a** Students should name any two gemstones, such as ruby, sapphire, emerald, amethyst.

**b** Diamonds are formed in hot igneous rock under intense heat and pressure.

**c** Gemstones can be expensive because they take millions of years to form and they are often located deep underground, so they are hard to extract. Rare gemstones are usually more expensive than those which are more common.

**6** For this extension question, students are asked to state the extent to which they agree with a statement. Good answers will give evidence to support both sides of the argument and provide a short conclusion.

An example answer: The rock cycle is important for humans because the food we eat grows due to minerals found in soil, which come from rock. The built environment which we inhabit is also a product of the rock cycle because buildings and roads are built using rock, concrete and cement. Metal ores are also found in rock; and we use metals for making cars, fridges, phones, etc.

However, there are other natural cycles that are essential too, such as the water cycle and the carbon cycle. Food could not grow without the humus in soil, which is made from rotting vegetation from plants – not a part of the rock cycle.

Overall, whilst we do rely on many natural processes to support the way we live, the rock cycle is important to a great extent because without it, we would have no food to eat or buildings to live in. We could possibly survive without buildings, but not without food!

# 1 Further suggestions for class and homework

Most of these suggestions are addressed to your students. Where research or further resources are needed, the Internet will almost certainly provide the answer.

The suggestions are graded *, **, *** according to the level of difficulty. Some are suitable for all levels, and can be differentiated by outcome.

## Our rocky home

1 **Rock walk** Walk around your local area, and look for rock. (Don't confuse it with concrete.) You could take photos and link them to an annotated sketch map of the area, on a wall display. Students could work in groups. */**/***

2 **Rock for fun** See how many outdoor activities you can come up with, that depend on rock. Find photos. */**/***

3 **Quartz** Quartz is a very common mineral in rocks, and has lots of uses. Which elements is quartz made from? What are its main uses? Find out, and tell us. **/***

## What are the three rock groups?

4 **The Giant's Causeway** What is it, and where? Mark and label it on a map, explain what it is made of, how it was formed, and add photos. */**/***

5 **Pumice** You'll find this igneous rock in a chemist shop. How was it formed, and what is it used for? **/***

6 **Marvellous marble** There isn't much marble in the UK – because most of our limestone deposits were not subjected to enough pressure. Find out why marble is so useful. Collect images and make a slideshow. ***

7 **Rock rap** Write a poem or rap about rock – or make up a children's story, where rock plays a big part. */**/***

## What is weathering?

8 **Choose one type of weathering** Choose one weathering process and do a presentation on it. Do some extra research first. Don't forget photos. **/***

9 **A weathering walk** Rock weathers. So does paintwork, timber, metal, and other substances exposed to the atmosphere. For example, paint flakes, and iron rusts. Walk around your school or local area, and look for signs of weathering. You could take photos and attach them to an annotated sketch map on a wall display. **/***

## What is the rock cycle?

10 **Go bigger** Make a big colourful version of the rock cycle for the classroom wall. Add notes and photos. Students could work in groups for this, sharing out the tasks. */**

11 **What's the connection ...** between limestone, chalk and marble? Write a paragraph about what links them, and add photos. **/***

12 **Your story** Once you were a grain of calcium carbonate in a sea shell. Now here you are, in Michelangelo's statue of David. Tell your story. */**/***

## The British Isles on their travels

13 **On the move** Five hundred and fifty million years ago, the land that forms the British Isles was in the southern hemisphere. Tell the story of its travels, from 550 mya to today, as a display on the classroom wall. You can base it on the globes on page 14 of the students' book. Beside each globe, name the geological period. Add notes and images to show life on Earth in that period. Students could work in teams, each team doing the research for one globe. **/***

14 **Remember the names of the periods** Create a mnemonic. For example: Come On, Show Dad Carlo's Present. Three Juicy Crabs! Price Nine Quid! ***

15 **Imagine ...** that Earth has only one big continent again, called Pangaea II. In what ways would our lives be different? Where on Pangaea II would you choose to live? */**/***

## Rock and relief around the UK

16 **The rock in your area** What kind of rock are you sitting on? Find out and tell the class. ***

17 **Volcanoes in the UK** The UK has a number of extinct volcanoes. Find out where they are and mark them on a map. Add photos if you can! ***

18 **Rock catalogue** Make a catalogue of rock types found in the UK. For each entry in your catalogue, say how the rock was formed, and add a photo. **/***

## What's the link between rock and landscape?

19 **Rock gallery** Using the map on page 17 of the students' book to help you, work your way around the UK and find photos of landscapes in areas of different rock types. ***

## How are *you* linked to soil?

20 **My life as an earthworm** Tell us about your life – where you live, what you eat. What can you see? */**/***

21 **It's elementary** Write a list of the elements in your body. Most of them come from soil! **/***

22 **Crossword puzzle** Make up a crossword puzzle based on this chapter. Give it to a partner to solve. */**/***

23 **Play Taboo** Prepare folded slips of paper, containing terms found in geology. A student chooses one and describes the term to the class, without using the term itself. How quickly will someone guess? You could set a time limit. **/***

# 2 Using Earth's resources

## About this chapter

### Addressing the KS3 Programme of Study

- Understand, through the use of detailed place-based exemplars at a variety of scales ...
  - human geography, relating to the use of natural resources
  - how human activity relies on effective functioning of natural systems

### About the content

- After introducing natural resources in general, this chapter focuses on three resource issues – water, food, and energy.

## The big picture

These are the key ideas in this chapter.

- Natural resources occur naturally in the environment – and we make use of them.
- Some natural resources are renewable and will not run out, no matter how much we use. For example, sunlight. Others are non-renewable. There may be a limited quantity, or we are using them faster than they are forming.

Earth's resources are not shared equally, and as the population rises, demand increases.

There are some solutions to resource issues, but we must learn to live in a sustainable way.

## The chapter outline

Use this to give students a mental roadmap for the chapter.

2 **Using Earth's resources** The chapter-opening photo in the students' book is an important resource. See opposite for using it.

2.1 **Earth's natural resources** What they are, and how they're not shared equally

2.2 **Water part 1 – Do we have enough?** Where water comes from, what we use it for, and water stress

2.3 **Water part 2 – Emptying the Ogallala?** How water is being pumped from aquifers, and the risk of running dry

2.4 **Water part 3 – Tackling water stress** Different approaches to avert a water crisis

2.5 **Food part 1 – Feeding the UK** Is our food security at risk?

2.6 **Food part 2 – Food insecurity, a global issue** Why is there so much food insecurity around the world?

2.7 **Food part 3 – Tackling food insecurity** Some approaches to tackle food insecurity

2.8 **Energy part 1 – Goodbye to the fossil fuels?** Is it time to switch to renewable sources of energy?

2.9 **Energy part 2 – Renewables in the UK** How electricity is generated from renewable sources in the UK

2.10 **Energy part 3 – Is solar power the solution?** How solar power can be used around the world

2.11 **Why must we change our ways?** How we are putting increasing pressure on Earth's resources, and how to live sustainably

2 **Using Earth's resources check questions** A page of questions to check knowledge, understanding, and skills across the chapter, with an opportunity for extended writing

## Objectives and outcomes for this chapter

| Objectives | Unit | Outcomes |
|---|---|---|
| Most students will understand: | | Most students will be able to: |
| • that we depend on natural resources | 2.1 | • define *natural resource, renewable resource* and *non-renewable resource* and give two examples of each |
| • that we depend on fresh water | 2.2 | • say what we use fresh water for |
| • that water stress is rising | 2.2 | • give reasons why water stress is increasing |
| • that water can become a non-renewable resource | 2.3 | • describe how and why water is being pumped from the Ogallala aquifer faster than nature can replenish it |
| • that we must take steps to tackle water stress | 2.4 | • give at least four examples of what could be done, to tackle water stress |
| • that the world's food security is at risk | 2.5 | • explain how population growth, politics and climate change can affect food security |
| • why so many people live with food insecurity | 2.6 | • explain the link between food insecurity and poverty |
| • that there are some ways to tackle food insecurity | 2.7 | • describe different ways to tackle food insecurity |
| • that the world still depends on fossil fuels | 2.8 | • describe the trend in demand for fossil fuels from 1800 |
| • how electricity is generated from renewable sources in the UK | 2.9 | • describe the distribution of solar and wind farms in the UK |
| • that solar power can be used around the world | 2.10 | • give the advantages of solar power, and decide which two renewable sources of energy could help to turn the world carbon neutral by 2060 |
| • that exploiting natural resources impacts other species | 2.11 | • explain how living sustainably will benefit future generations |

These tie in with the outcomes for each unit in this teacher's handbook.

## Assessment for this chapter

There are assessment materials for this chapter on *geog.3 Kerboodle Lessons, Resources, & Assessment*. They include interactive auto-marked end-of-lesson and chapter-level tests, a printable exam-style end-of-chapter paper (with a mark scheme), and a printable self-assessment checklist. See also the notes on pages 14–17 of this book.

## Getting ready for this chapter

There is a scheme of work for this chapter, plus a glossary and glossary worksheet, on *geog.3 Kerboodle Lessons, Resources, & Assessment*.

## About the chapter-opening photo

Ask students:

Could we run out of water to grow crops like these potatoes?

What will happen if soil is over-used?

The image shows two potato farmers in Sacred Valley, Cuzco, Peru – in the Andes. Potatoes were first cultivated by the Incas (8000–5000 BCE). They did not reach Europe until the 16th century. Now they are a staple food in many places around the world.

# 2.1 Earth's natural resources

Help at a glance

## About this unit

This opening unit shows how we depend on Earth's natural resources to meet our needs, and then presents some key definitions.

## Key points

- Humans have always depended on Earth's natural resources, and continue to do so.
- A natural resource is something which occurs naturally, and which we make use of. For example, soil, water, wind.
- Some resources are renewable – they won't run out, no matter how much we use them. For example, sunlight. Others are non-renewable.
- Natural resources are not distributed equally around Earth. For example, some places receive little rain, so may have a scarcity of fresh water. Some places receive plenty.

## Key vocabulary

finite, natural resource, renewable resource, non-renewable resource

## Skills practised in 'Your turn'

- Literacy skills: q1, q3, define; q5, describe
- Thinking skills: q2, give examples; q2, q4, decide, and explain; q5, come up with a survival plan

## Unit outcomes

By the end of this unit most students should be able to:

- explain the terms given in 'Key vocabulary' above
- explain clearly the difference between a natural resource and a finite resource; a renewable and non-renewable resource
- say that natural resources are not distributed equally across Earth, and give at least one example
- say that we make use of other countries' natural resources, for example when we buy food from other countries

## Ideas for a starter

**1** With books closed the students should play a mind movie. Ask students to close their eyes and think back 12 000 years, to the time when people were returning here, at the end of the ice age. What is it like for them? What will they need, to survive? Where will they get it? Create a spider map on the board.

**2** With books closed, tell the class that you are about to show them a set of images. The competition is to see if they can identify the theme (*using Earth's natural resources*) before the end of the set. Then display the images. (Choose at least some without people, to make it clear that the theme is not *employment*.) For example; water pouring from a tap, a field being ploughed (with the focus on the soil), gas being burned by a gas cooker, trees being felled for timber (in the UK), quarrying, a field of solar panels.

## Ideas for plenaries

**1** Any or all of questions 1–4 of 'Your turn' could be used as plenaries.

**2** Ask: Which do you think it's better to have: renewable resources or non-renewable ones? Why?

3 What natural resources do we have in this area, right now? (Rain, river, sunlight, air, soil, rock …)

4 What could you do without, if you had to? Prompt students to keep whittling away their lists until only the basics are left. Students could work in pairs for this.

## Further class and homework opportunities

Suggestions 1 and 2 on page 63 of this book

*geog.3 workbook*, page 12

**kerboodle**

### Resources

2.1 You decide! – Students rank the importance of natural resources.

2.1 Sort it out! – Students categorise renewable and non-renewable resources.

2.1 Foundation worksheet – Students consider the difference between need and want, list the problems with Earth's resources, and identify natural and renewable resources.

2.1 Extension worksheet – Students analyse natural resources, their importance and associated problems.

### Assessment

2.1 End-of-lesson test – An assignable, auto-marked formative assessment.

## Answers to 'Your turn'

**1 a** A natural resource is one which occurs naturally in the environment, and which we can make use of; for example, wind and oil.

**b** A finite resource is one which there is a limited amount of, so it could run out one day.

**2 a, b**

**A** Yes, it occurs everywhere. We breathe oxygen from it, and depend on crops taking carbon dioxide from it; we use it for pumping things up (footballs, balloons, air beds, bicycle tyres); we extract other gases from it too; for example nitrogen, which is used to fill food packaging to keep things fresher for longer, and to preserve human tissue (when used as liquid nitrogen); and helium, neon, and other noble gases (which we use in lighting).

**B** No; it is made in factories (usually from natural materials).

**C** Yes; we live on top of rock and it is all around us. For ways we use it, see the answer for question **5**, on page 21 of this book.

**D** Yes, it occurs naturally. We depend on it for our supply of fresh water, and to water crops.

**E** Yes, oceans occur naturally. We travel on oceans, and use them as a source of tidal and wave energy, and for fun (sailing, swimming, and so on).

**F** No; it is made in factories (usually from natural materials).

**G** Yes, it occurs naturally. We use rainforest plants in medical research; many of our foodstuffs originated in the rainforest; environmental tourism into rainforest areas is popular. Some people still illegally log timber from rainforests, and burn them down for farmland. Some people live sustainably in the rainforest, by hunting, and gathering plants, fruits, and seeds.

**H** No; it is made in factories (mostly from natural materials, but it may also contain artificial flavourings and sweetener).

**3 a** a resource which nature continues to provide, or which we can keep growing more of; for example, rain or wood

**b** one which we are using up, and which may run out one day

**4 a** renewable; wind occurs naturally as a result of differences in air pressure

**b** non-renewable; coal originated in the Carboniferous period, when lush swamps died off and were buried; perhaps small quantities are still forming somewhere, but we are using it up much faster than it can form

**c** non-renewable; there is only so much gold within Earth

**d** renewable – as long as there is sunshine, and bodies of water

**5** The base of one glass bottle could be used as a magnifying glass, for starting fires. But you'd probably have to break the bottle.

# 2.2 Water part 1 – Do we have enough?

Help at a glance

## About this unit

This unit explores where we get our fresh water from, what we use it for, and whether we have enough of it.

## Key points

- Less than 1% of the Earth's total water supply is fresh water which we can use.
- In the UK around 28% of our water supply is groundwater, the rest is surface water.
- Globally, around 70% of water is used for farming, 19% for industry, and 11% for is used in homes. This pattern varies around the world.
- Demand for fresh water is growing, but climate change and population growth are resulting in water stress.
- Water stress is where a country cannot meet its demand for fresh water.

## Key vocabulary

salty water, fresh water, groundwater, aquifers, reservoirs, water vapour, water stress

## Skills practised in 'Your turn'

- Geographical skills: q2, q5, interpret a diagram; q6, compare maps; q6, q7, interpret a map
- Literacy skills: q4, write sentences to correct a false statement; q5, define
- Thinking skills: q1, explain; q3, suggest issues; q5, q7, give reasons

## Unit outcomes

By the end of this unit most students should be able to:

- explain the terms given in 'Key vocabulary' above
- explain how nature turns salt water from the oceans into fresh water
- describe the global use of fresh water
- give two reasons why water stress is growing
- name four countries that may suffer extremely high water stress by 2040

## Ideas for a starter

1. Show an image of Earth from space which also shows ice. Ask: How much of the world's water can we use? How much is salty water? How much is locked up in ice and glaciers? How much is underground?
2. With books closed ask: What do we use fresh water for? Draw a spider diagram on the board. Group them together – domestic, industrial, and farming uses. Ask which category uses the most water.

## Ideas for plenaries

1. Discuss map D on page 27 of the students' book. What patterns can students see? Why is the UK likely to suffer from medium to high water stress by 2040?
2. Ask students to write the term FRESHWATER vertically down their page. It is the spine down which students write a key phrase or term from today's lesson. Each phrase or term must contain the given letter – but not necessarily at the start. For example, for S, the term could be *water Stress*.

## Further class and homework opportunities

Suggestions 3 and 4 on page 63 of this book

*geog.3 workbook*, page 13

**kerboodle**

### Resources

2.2 Sort it out! – Students categorise the uses of water.

2.2 Where in the world? – Students interpret a map of water stress.

2.2 Foundation worksheet – Students describe how they use water and how its loss would affect them, and fill in the gaps of a paragraph outlining the problems of accessing water.

2.2 Extension worksheet – Students answer and write their own true or false statements about water, and use a map to identify countries likely to experience high water stress.

### Assessment

2.2 End-of-lesson test – An assignable, auto-marked formative assessment.

## Answers to 'Your turn'

**1** The key point is that when water evaporates from the ocean, the salts are left behind. So clouds and rain contain no salts. The rainwater then feeds rivers and aquifers. The fresh water in these will contain some salts dissolved from river banks, and soil, but our bodies are able to cope with their low concentrations.

**2** Less than 1% is available for us to drink (3% of Earth's water is fresh so we *could* drink it – but some is trapped in ice and glaciers so is not accessible).

**3 a i** Groundwater collects below the ground, when rain trickles down through the soil.

**ii** An aquifer is an underground rock structure, which holds groundwater.

**b** Some aquifers may be very deep; the pump needs to be powered, which requires infrastructure; some impoverished regions may not have the technology; conflict between countries which share an aquifer.

**4** False.

Students should write sentences such as:

- 'Very little of the fresh water on Earth is used for drinking.'
- 'Most of Earth's fresh water is used for agriculture.'

(70% of fresh water is used for agriculture; 19% is used in industry; and only 11% is for domestic use, which includes washing, cooking, etc.)

**5 a** Water stress is where a country can't meet its demand for water.

**b** Two from: rain falls unevenly around the world so some places have less fresh water than others; climate change means some places are getting even less rain than before; if aquifers or rivers run dry there may be a water shortage; Earth's population is growing quickly so demand for water is high.

**6 a** Students should name four countries with high water stress such as: Saudi Arabia, Iran, Kazakhstan, Spain. Countries are likely to be from North Africa or western Asia, including the Middle East.

**b** Students should name four countries with low water stress such as: Brazil, Colombia, Ethiopia, Vietnam. These are likely to be from South America, central Africa, or South East Asia.

**7 a** Medium to high water stress

**b** Water stress will vary across the UK because it depends on factors such as rainfall levels and population density. The level of water stress is calculated by comparing how much is extracted to how much is available. The south-east will experience greater stress because it has a higher population (London) and a drier climate than the north and west.

# Water part 2 – Emptying the Ogallala?



## About this unit

This unit is about the Ogallala aquifer in the USA, where water is in effect a non-renewable resource. It is being pumped out of the aquifer much faster than it can be replenished.

## Key points

- The Ogallala aquifer below the USA's Great Plains is a huge aquifer. Eight states pump water from it.
- The water in the Ogallala is mostly fossil water – ancient water trapped from the last ice age.
- The water level is falling, it is being pumped out faster than it can be replenished.
- If the level keeps falling, many people will be affected – and in particular the farmers.
- The Ogallala is not the only large aquifer being depleted. The Nubian Sandstone aquifer below the Sahara is another example.

## Key vocabulary

aquifer, fossil water, breadbasket, Nubian Sandstone aquifer, Great Man-made River

## Skills practised in 'Your turn'

- Geographical skills: q1, q2, interpret a map; q2, interpret a diagram; interpret a photo, q3, q4, q5
- Literacy skills: q2, define; q8, justify
- Thinking skills: q1, describe; q2, q3, q4, explain; q5, suggest differences; q6, list; q7, q8, decide and explain

## Unit outcomes

By the end of this unit most students should be able to:

- explain the terms given in 'Key vocabulary' above
- say what and where the Ogallala is
- explain why the water levels in the Ogallala are falling, and describe the likely consequences
- say why it would take thousands of years for the aquifer to be replenished

## Ideas for a starter

1. With books closed, write OGALLALA on the board. By asking you 20 yes / no questions, students have to guess what and where it is. Give the class a starting clue by saying that it's a natural physical feature.
2. With books closed, ask: What's an aquifer? (Met in *geog.1*!) Students have to answer by silently drawing a diagram on the board (or in exercise books, working in pairs). Prompt for more. For example: Where did the water come from? How do we get at it?

## Ideas for plenaries

1. Discuss diagram E. Which state has the smallest share of the Ogallala? Which one has the least water left below it? Is there a limit to how deep a well you could dig, and still find water?
2. So what would you do about pumping water, if you were in charge of protecting the Ogalalla aquifer?
3. What is the big key point behind today's lesson, which would apply anywhere in the world? Turn it into a slogan for an environmental campaign.

## Further class and homework opportunities

Suggestions 5 and 6 on page 63 of this book

*geog.3 workbook,* page 14

**kerboodle**

### Resources

2.3 Case study: The Ogallala aquifer – Student create a case study of the Ogallala aquifer.

2.3 Quiz time! – Students complete a quick quiz about the Nubian Sandstone aquifer.

2.3 Foundation worksheet – Students consider the finite nature of the Ogallala aquifer and write a story explaining what the situation might look like in 20 years.

2.3 Extension worksheet – Students match up numbers to facts on the Ogallala aquifer and assess ways of managing water shortage and crop failure.

### Assessment

2.3 End-of-lesson test – An assignable, auto-marked formative assessment.

## Answers to 'Your turn'

**1 a** The Ogallala is a huge aquifer which lies below the Great Plains. The Great Plains extend from Canada down through the central states of the USA – but the aquifer is only in the USA.

**b** Any four from: South Dakota, Wyoming, Nebraska, Kansas, Oklahoma, Texas, New Mexico, Colorado.

**c** It is so important to farmers because they pump water from it to irrigate their crops. Without this water their crops would not grow, and they may not be able to make a living from farming.

**2 a** Fossil water is water that has been sealed in aquifers for thousands or millions of years.

**b** The level of water in the Ogallala aquifer is falling because it is being pumped out in large quantities for agriculture but is not replenished, as there is not much rain in the region.

**c** The level is lowest under Texas.

**3** They are clouds of dust. They formed because the soil was very dry as a result of drought, and the wind was able to whip up the dust from it.

**4** There are big green circles because crops are planted in circles, so that they can be irrigated by a central-pivot irrigation system. The brown parts are not planted or irrigated.

**5** Fifty years from now, if the Ogallala has run out of water, you might expect to see no circles, and no rotating sprinkler arms. There might be less-thirsty crops growing, that rely on rainfall alone – or perhaps the land will have been abandoned. If pumping from the Ogallala has been controlled, there may still be green circles, but perhaps fewer.

**6** The groups could include: farmers and their families; local shops and other services which rely on the farmers for income; local food-processing factories; shoppers around the USA who may have to pay more for bread and other foodstuffs, since the supply of crops from the Ogallala region has fallen away; factories and shoppers in other countries, because the USA may have less foodstuff available for export; companies who transport and ship crops and foodstuff; people in poorer countries who may receive less food aid from the USA; the governments of the Ogallala states, who will receive less in taxes from farmers and food-related companies, and will have to pay out more in benefits to the people who have lost their jobs.

**7** True. The water in the Nubian Sandstone aquifer is fossil water. Students should point out that the Sahara is a very arid region with very little rainfall. Despite these conditions a vast aquifer (2.6 million sq km) has formed below it. The aquifer's size combined with the region's low rainfall levels indicate that it most likely formed in the past.

**8 a** Egypt, Libya, Sudan, Chad

**b** Yes. Water causes conflict because it is a resource which is scarce and / or running out in some places.

# Water part 3 – Tackling water stress



## About this unit

This unit looks at a variety of approaches to meeting the demand for fresh water.

## Key points

There are a number of approaches that can be used to meet the demand for fresh water.

- Harvesting the rain – whether it's in simple containers or harvesting ponds.
- Reducing waste – for example by fixing leaks in the supply, using drip irrigation, and changing the type of crops grown to those that need less water.
- Recycling used water – for farming, or even for drinking.
- Moving it from areas of surplus to areas of deficit, for example China's South–North Water Transfer Project.
- Desalinating sea water.

## Key vocabulary

harvesting ponds, drip irrigation, South–North Water Transfer Project, desalination (and desalination plants)

## Skills practised in 'Your turn'

- Geographical skills: q4, interpret and compare maps
- Literacy skills: q1, summarise
- Thinking skills: q2, q6, explain; q2, q3, suggest; q3, evaluate and explain; q5, assess; q7, discuss

## Unit outcomes

By the end of this unit most students should be able to:

- explain the terms given in 'Key vocabulary' above
- identify which approaches to meeting the demand for fresh water could be implemented in the UK
- identify which approach might impact the environment the most
- name four countries facing extreme water stress that could make use of desalination

## Ideas for a starter

1 Hold up a glass of water. Say this lesson is about meeting the demand for what is in the glass. Give the class x minutes (as long as you think is needed) to think up three key questions to ask on this topic. Students could discuss the questions in small groups and the class could vote for the best three. Note the questions on the board. You can return to them later.

2 Ask: How do we waste water? Record students' suggestions on a spider map. Can students recall the biggest user of water (agriculture)? It's also where most waste happens.

## Ideas for plenaries

1 If you chose starter 1, revisit the questions on the board. Which have been answered during the lesson? Which haven't? The latter could become homework projects.

2 Ask: Why is there a desalination plant on the Thames?

3 Give students 4 or 5 minutes to summarise what they have learned about water as a resource. They could work in pairs to create a mind map and add drawings.

## Further class and homework opportunities

Suggestions 7 and 8 on page 63 of this book

*geog.3 workbook,* page 15

**kerboodle**

### Resources

2.4 You decide! – Students rank the potential solutions to water stress.

2.4 Where in the world? – Students identify countries which could use desalination.

2.4 Foundation worksheet – Students design a leaflet for farmers to inform them of how they can cut down on water use.

2.4 Extension worksheet – Students use pages 30 and 31 of *geog.3* to complete a table on tackling water stress, outlining approaches and their advantages and disadvantages.

### Assessment

2.4 End-of-lesson test – An assignable, auto-marked formative assessment.

## Answers to 'Your turn'

**1** Students should briefly summarise each of the five boxes.

**2** **a** The two methods which would cost the least are: 'stop wasting it,' and 'catch it as it falls.' 'Stop wasting it' would cost nothing and would actually save water. The only cost for the 'catch it as it falls' method is the containers used to catch the water. However, if farm workers are being paid a wage there may also be an associated cost of the time taken for them to carry the water.

**b** Most likely to be implemented in the UK: 'stop wasting it' – people could be encouraged to use less water and leaks could be fixed ; 'recycle it' – the UK already has sewage works which do this job, especially in urban areas where water is scarce; 'take it from the sea' could be implemented as we are surrounded by the ocean and many of the UK's largest cities are near the sea.

**c** A landlocked country with little rain could save water by using the 'stop wasting it' method, or build sewage works to 'recycle it.'

**d** 'Stop wasting it' would have the most positive effect on the environment; whereas 'move it' would probably have the most negative effect. To move water on a mass scale requires a huge amount of infrastructure (roads, canals) and building these uses energy and resources.

**3** China is willing to spend so much money moving water because the north is dry, but this is where most of the agricultural land is located. The north needs water to grow food to feed China's large population. Students may also mention that the capital city, Beijing (population 22 million), is in the north.

**4** Students should name four countries with extreme water stress that have a coastline (this includes those that border inland seas). For example: Portugal, Spain, Morocco, Algeria, Libya, Saudi Arabia, Yemen, Turkey, Iran, Pakistan.

**5** 'Stop wasting it' would be the best solution overall. This logic can be applied to all of Earth's resources. If we use less, our lives will be more sustainable.

**6** In fact people will probably move of their own accord, where water shortage is a serious issue, because drought may mean death. The question is: where can they go? By moving, they will also increase the demand for water somewhere else, where the supply may not be able to cope. (For example in a refugee camp.) This is a difficult problem facing the world.

**7** To discuss, students should look at something from different angles and give key points about it. On one hand, these methods could save the world from a severe water crisis because most countries could utilise at least two of these methods. On the other hand, there are other solutions not mentioned here, such as protecting wetlands and addressing pollution of fresh water sources. Overall, if the population continues to rise and wealth keeps growing, people will continue to need more water, and it may be difficult to prevent a crisis.

# 2.5 Food part 1 – Feeding the UK

Help at a glance

## About this unit

This unit considers whether there is a risk of the UK running out of food, now or in future.

## Key points

- The UK has a high level of food security – where people have access to sufficient safe and nutritious food.
- However, not everyone in the UK can afford enough food, and they struggle with food insecurity.
- We are not self-sufficient in food – about 55% of the UK's food is produced here.
- The UK's food security is at risk due to population growth, politics and climate change.
- Experts think that more food should be produced in the UK, so that we become more self-sufficient.

## Key vocabulary

food security, food insecurity, self-sufficient, trade deals, European Union (EU), tariffs, trade wars, greenhouses, hydroponics

## Skills practised in 'Your turn'

- Geographical skills: q6, interpret photos
- Literacy skills: q1, q2, define
- Thinking skills: q2, q5, give reasons; q3, suggest; q4, explain; q5, q7, give an opinion; q6, identify differences; q7, justify; q7, list advantages

## Unit outcomes

By the end of this unit most students should be able to:

- explain the terms given in 'Key vocabulary' above
- give one reason why the number of people using food banks in the UK increased during the COVID-19 pandemic
- give examples of how population growth, politics and climate change can affect food security
- give three examples of ways that more food could be produced in the UK in future

## Ideas for a starter

1. Food security – ask students what they think this term means. What is food insecurity? Be aware that this could be a contentious, and sensitive, issue for some of your students.
2. Show students photo D from page 33 of the students' book. Ask them what is growing here? How is it growing? Is this how we grow most of our food in the UK? Will we need to do more of this in the future?

## Ideas for plenaries

1. Ask: Would you eat meat grown in a lab? Why? Why not? Have a class discussion.
2. Will the UK run out of food? Ask students to work in pairs to create a spider map in response to this question, based on what they have learned in this lesson.

## Further class and homework opportunities

Suggestion 9 on page 63 of this book

*geog.3 workbook,* page 16

**kerboodle**

### Resources

2.5 Word scramble – Students complete a crossword about food production in the UK.

2.5 Quiz time! – Students complete a quick quiz about food security in the UK.

2.5 Foundation worksheet – Students answer questions on the UK's food security and self-sufficiency, and evaluate how the UK could improve these in the future.

2.5 Extension worksheet – Students write a speech on the UK's food security, explaining why it is at risk and how we can mitigate this problem in the future.

### Assessment

2.5 End-of-lesson test – An assignable, auto-marked formative assessment.

## Answers to 'Your turn'

**1** **a** Food security is when people have access to enough safe and nutritious food for a healthy and active life.

**b** Food insecurity is the opposite. It affects people who cannot afford enough food to meet their needs.

**c** Self-sufficiency means producing enough food to meet requirements. A country could be self-sufficient if it doesn't need to import any food. People can be self-sufficient if they grow all their own food.

**2** **a** A food bank is a place which gives out free (donated) food to people who need it.

**b** Many people lost income so they couldn't afford food. Reasons include: some people lost their jobs; some were unable to work due to illness; some businesses had to close during lockdowns.

**3** Importing food means the UK has more variety, for example, bananas and cocoa are not grown in the UK. In the winter, foods which only grow in the UK in summer are still available, such as strawberries. Imported food is often cheaper than food produced here.

**4** **a** As Earth's population grows, there is more competition for imports of food, which could affect UK food security if there are supply issues or if prices rise.

**b** High tariffs on imports causes food to become more expensive. If people in the UK cannot afford the higher prices, it could affect food security.

**5** Students should agree that climate change could increase food insecurity in the UK. Reasons: other countries may have less food to sell; British farmers may be affected by heatwaves and floods; new pests and diseases may become more common.

**6** **a** Differences: B is growing outside, whereas D is inside. The plants in B receive sunlight, but in D artificial light is used. More land is required for B, whereas plants in D are grown vertically in layers. Soil is used in B, but the plants in D are fed nutrients without the need for soil.

**b** Hydroponics

**c** D – Hydroponics because this method requires less space.

**7** **a** Students should give their own opinion on lab-grown meat. Some may wish to eat the chicken bites for some of the reasons listed in part b, below. Some students may be against it on the grounds that this is not a natural method of producing food; or because farmers may lose their jobs. This question would make a good class debate.

**b** Advantages: if more people ate lab-grown meat, it would reduce the need for large scale factory farming; no animals are harmed in its production; fewer greenhouse gases are associated with lab-grown meat so it would help tackle climate change; agricultural land could be reclaimed and used for other things, or rewilded.

# 2.6 Food part 2 – Food insecurity, a global issue

Help at a glance

## About this unit

In this unit students will explore why approximately 2 billion people live with food insecurity.

## Key points

- Many countries (approximately 2 billon people) have high levels of hunger and food insecurity.
- Africa is the continent with the highest levels of food insecurity.
- Poverty is a key factor in food insecurity, but pests and diseases, conflict, climate change, and water stress also play a part.
- Food insecurity can have a wide range of impacts, and can tip over into famine, resulting in crisis.

## Key vocabulary

food-insecure

## Skills practised in 'Your turn'

- Geographical skills: q2, interpret choropleth map
- Numeracy skills: q2, interpret percentages using a map key
- Literacy skills: q2, comment
- Thinking skills: q1, q3, q4, explain a link; q3, q6, give reasons; q4, make suggestions and justify; q6, imagine a scenario

## Unit outcomes

By the end of this unit most students should be able to:

- explain the term given in 'Key vocabulary' above
- name the continent, and three countries within it, with the highest percentage of people living in hunger
- be aware that there is some food insecurity in every country
- list the factors that play a part in food insecurity
- give three impacts of food insecurity

## Ideas for a starter

1. Ask a student to read out 'What's for dinner?' on page 34 of the students' book, to set the scene for this lesson.
2. Show students map A from page 34 of the students' book on the board. Can they recall the definition of food insecurity? Can they describe the pattern shown on the map?
3. With books closed, write FOOD INSECURITY on the board. Say that this is today's topic. Ask for four questions which the lesson might answer, related to food insecurity, beginning with *What… Where… Why… Who…*

## Ideas for plenaries

1. Use 'Your turn' question 4 as a plenary activity.
2. Choose a student to be in the 'hot seat'. Other students can fire sensible questions at them on today's lesson. When they can't answer, another student takes over.
3. End poverty, and you will sort out food insecurity. Discuss.

## Further class and homework opportunities

Suggestion 10 on page 63 of this book

*geog.3 workbook,* page 17

**kerboodle**

### Resources

2.6 You decide! – Students rank the causes of food insecurity.

2.6 Where in the world? – Students interpret a map of world hunger.

2.6 Foundation worksheet – Students use the map on page 34 of *geog.3* to locate Malawi and write a diary entry analysing the impact of food insecurity.

2.6 Extension worksheet – Students evaluate whether human or natural causes are the biggest cause of food insecurity and fill out a diagram on the cycle of poverty.

### Assessment

2.6 End-of-lesson test – An assignable, auto-marked formative assessment.

## Answers to 'Your turn'

**1** Yes. Hunger is a strong indicator of food insecurity because it relates to lack of food.

**2** **a** Africa

**b** Any three from: Chad, Central African Republic, South Sudan, Congo, Uganda, Somalia, Madagascar, Liberia.

**c** In Malawi, 15–25% of the population live in hunger.

**3** **a** One from: because she doesn't have money for fertilisers; because she can't afford new farming tools; because her husband died so she has to tend the crop alone.

**b** One from: because the rain stopped early, which may have been an effect of climate change; because climate change has made rainfall more irregular; because other issues may affect her crop, such as floods or pests.

**4** **a** Food insecurity is linked to water because if there is a lack of water, farmers may not be able to irrigate their land and may struggle to grow healthy food crops.

**b** Food insecurity is linked to roads because if there are poor roads, it may be hard to transport food to market.

**c** Food insecurity is linked to conflict because when there is conflict, transport routes may be blocked, so food cannot be moved. In extreme cases, farmers may have to flee and abandon their farms.

**d** Food insecurity is linked to soil because good crops grow in nutrient-rich soil. If farmers can't afford fertiliser to enrich the soil, the yields may be poor.

**e** Food insecurity is linked to crop pests because some crops are completely destroyed by pests such as locusts. In places where farmers can't afford pesticides, this is very common.

**5** **a** Any points from 1 to 4 would be the best suggestions here. For example, the government could improve roads, provide training for farmers and subsidise fertiliser / pesticides.

**b** Many of these points could be justified. For example, poverty may be reduced with foreign aid or cancelling of debts, or with help from other countries to remove corruption from the government. However, students are likely to focus on climate change as it is a global issue that is mostly created by emissions from developed countries. Actions such as the 2015 Paris Agreement are required to solve complex global issues.

**6** Students should write about two other factors here, which were not previously mentioned in their answer to question 5. Chifundo is likely to want her government to focus on the issues which have the most obvious effects on her daily life, so poverty will be a common theme.

# 2.7 Food part 3 – Tackling food insecurity

Help at a glance

## About this unit

This unit looks at some of the ways food insecurity is being tackled in Malawi and Togo.

## Key points

- NGOs have helped some farmers to improve food security in Malawi by providing hybrid seeds, fertiliser, and storage sacks.
- They also provide help with farming techniques, including planting in pits to trap and store rainwater and microdosing fertiliser.
- In Togo, technology helped to locate the poorest people during the COVID-19 pandemic, and they were sent vouchers on their mobile phones to use to buy food or get cash.
- Satellites, drones, and mobile phones are all helping to tackle food insecurity.

## Key vocabulary

There is no new key vocabulary in this unit.

## Skills practised in 'Your turn'

- Geographical skills: q4, interpret photos
- Literacy skills: q2, q3, define
- Thinking skills: q1, identify benefits; q2, provide an opinion; q2, q3, explain; q4, imagine a scenario; q4, make suggestions; q5 make links between content, discuss various ideas

## Unit outcomes

By the end of this unit most students should be able to:

- describe how NGOs have helped some farmers in Malawi
- describe some of the ways that farmers in Malawi have benefitted from being able to sell food
- explain how satellites and mobile phones helped to tackle food insecurity in Togo during the COVID-19 pandemic
- explain how technology can help to track the spread of new diseases affecting plants

## Ideas for a starter

1. Ask a student to read the text on 'Joy!' aloud to the rest of the class. As they are reading, ask other students to note down the ways that the NGO helped, and how Chifundo and her family will benefit.
2. Recap Unit 2.6. Where is hunger and food insecurity greatest? What issues does Chifundo face in Malawi?

## Ideas for plenaries

1. Discuss the photos in this unit. What is going on here?
2. Look at the method of planting in the lower right photo on page 36 of the students' book. What are the benefits of this method of planting?
3. Use 'Your turn' question 5 as a plenary activity.

## Further class and homework opportunities

Suggestion 11 on page 63 of this book

*geog.3 workbook*, page 18

kerboodle

### Resources

2.7 Quiz time! – Students identify the definition of an NGO.

2.7 You decide! – Students rank the solutions to food insecurity.

2.7 Foundation worksheet – Students write a continuation of Chifundo's diary entry from page 36 of *geog.3*, to explain the impact of a successful crop on her life.

2.7 Extension worksheet – Students consider strategies to tackle food insecurity and explain how food security can benefit a country as a whole.

### Assessment

2.7 End-of-lesson test – An assignable, auto-marked formative assessment.

## Answers to 'Your turn'

**1** Answers are likely to focus on how Chifundo could benefit from the profits made from selling food. For example: to buy chickens which will lay eggs for her family to eat; she could support one or more of her children through education; she could pay back her loan to the NGO.

**2** **a** An NGO is a charity (such as OXFAM) which helps people and is independent of the government.

**b** Students must consider an ethical dilemma and explain their stance. Some may say that the NGO would be more charitable if it provided things free of charge to people who are living in extreme poverty. Others might think it is fair for Chifundo to pay back her loan, because she will be able to make a profit from selling the food she has grown. As charities have limited funds, they will be able to help more people if loans are paid back. This would work well as a class debate.

**3** **a** Satellites can take images of large areas of farmland and the information they gather can be relayed to farmers to help them protect their crops. For example, the images can be used to analyse the direction and speed at which plant diseases may be spreading. Satellites can also use sensors to detect moisture levels in crops and soil quality, which would be useful information for farmers as it could help them make decisions on where to plant their crops.

**b** Artificial intelligence is when machines are programmed to do tasks that usually need human intelligence.

**4** Spider maps could include: hybrid seeds might improve the quality of the crop and / or make it resistant to drought; storing maize in specially designed sacks will make it last longer; planting crops in pits which trap rain will help with irrigation; sprinkling fertiliser around individual plants would mean less is needed; information from satellites may help make farming decisions; it may be possible to ask for more help from NGOs.

**5** To 'discuss,' students should look at something from different angles and give key points about it. To save the whole world from food insecurity would require co-operation on a global scale. Ideas from the unit may help. But other factors might come into play: conflict (wars) might interrupt supply chains or force people to leave their land; climate change might cause damaging floods or drought. So there are events or forces that might make it harder for some of the ideas to have an impact. Students might conclude that the ideas met in the unit might not be capable of saving the whole world from food insecurity in isolation.

# Energy part 1 – Goodbye to the fossil fuels?



## About this unit

This unit looks at how the world depends on fossil fuels, and why we need to switch to renewable sources of energy.

## Key points

- Fossil fuels (coal, oil, and natural gas) are non-renewable resources. They are used for generating electricity, for transport, for heating, and cooking.
- The use of coal has grown since 1800, and use of oil and natural gas grew rapidly in the 20th century.
- Fossil fuels release carbon dioxide when they are burnt, resulting in climate change.
- The world relied on fossil fuels for over 80% of its energy in 2019, but many countries aim to be carbon neutral by 2050.
- The world is switching to producing electricity from renewable sources.

## Key vocabulary

coal, oil, gas, non-renewable resource, wood fuel, greenhouse gas, Paris Agreement, carbon neutral, renewable energy sources, green electricity

## Skills practised in 'Your turn'

- Geographical skills: q2, q3, interpret and analyse a line graph; q6 interpret and analyse a divided bar chart
- Literacy skills: q1, unscramble anagrams; q6, define
- Thinking skills: q2, q4, comprehension; q2, q3, q5, explain; q2, q4, suggest a reason; q5, consider personal impacts; q6, comment on an issue

## Unit outcomes

By the end of this unit most students should be able to:

- explain the terms given in 'Key vocabulary' above
- describe the trend in the use of fossil fuels since 1800
- explain why so many people still depend on wood for fuel
- explain why many countries signed the Paris Agreement in 2015
- comment on the challenge the world faces in going carbon neutral

## Ideas for a starter

1. With books closed, say: It is now (*give the time*). How much have you depended on fossil fuels today? What have you used them for? Write students' responses on the board.
2. With books closed, show students the photo of the child carrying the basket of firewood on page 38 of the students' book. Ask: What is in the basket? Is it a fossil fuel? Is it renewable (yes, but over 2 billion people rely on firewood and charcoal for cooking and heating, so forests are disappearing)?

## Ideas for plenaries

1. Ask: Overall, have fossil fuels been a good thing or a bad thing? Discuss as a class.
2. How might this lesson have been different if it had been taught by an executive from an oil company? Or by the child carrying a basket of firewood?

## Further class and homework opportunities

Suggestions 12 and 13 on page 63 of this book

*geog.3 workbook,* page 19

kerboodle

### Resources

2.8 You decide! – Students rank the reasons for the increased use of fossil fuels.

2.8 Quiz time! – Students complete the explanation of fossil fuel use over time.

2.8 Foundation worksheet – Students define key terms, list the ways we use fossil fuels, and complete a paragraph explaining the problems of this use.

2.8 Extension worksheet – Students write a letter explaining the problems caused by fossil fuels and colour in a pie chart displaying where the world gets its energy from.

### Assessment

2.8 End-of-lesson test – An assignable, auto-marked formative assessment.

## Answers to 'Your turn'

**1** Burn; energy

**2 a** There is an upward trend of increasing use for all three fossil fuels. Before 1950, this was a slow steady increase; but from 1960 onwards all three experienced much faster growth.

**b** Oil

**c** Oil did not drive the Industrial Revolution. Students should note that the growth of industry began much sooner than oil became a popular fuel (1910). The graph shows that coal was rapidly gaining popularity in the mid-1800s when the Industrial Revolution was in full swing.

**3 a** Two from: petrol, diesel, jet fuel

**b** From 1908, oil consumption increased because it was used to make petrol for cars to use as fuel. (The Model-T Ford was the first car which was mass produced and sold at a price which made it accessible to large numbers of people.) From Graph A, it is evident that there was growth in the amount of oil used from around 1910.

**c** Demand for oil fell in 2020 due to the COVID-19 pandemic. Global travel virtually stopped as borders were closed, and local travel declined as a result of lockdowns across the world. Many industries closed for a period. These factors reduced the demand for oil.

**4 a** Logs, firewood, charcoal

**b** Many people still depend on wood fuel because they live in places which don't have the infrastructure (pipes) required to use electricity or gas; particularly those living in rural areas in poorer countries.

**5 a** The Paris Agreement was a response to climate change. Climate change is a result of rising levels of carbon dioxide (and other gases) in the atmosphere. Evidence suggests that this is closely linked to the burning of fossil fuels.

**b** The world must cut back sharply on using fossil fuels because as Earth gets warmer, climates are changing. Some regions will experience more droughts; others experience more floods. These changes will impact farming and make it harder to grow crops in certain places, which could lead to food insecurity. Droughts could lead to water insecurity. Warming temperatures are also causing Arctic ice to melt, and sea levels are rising which creates a flood risk for low-lying regions.

**c** The Paris Agreement is being acted upon differently in every country. Students may comment that taxes on fuel may go up, so car and air travel could become more expensive; more people may start to drive electric vehicles; initiatives to encourage people to walk or cycle to school may become more common; people will be encouraged to use products with less packaging to reduce the amount of energy used in production; people may start eating more locally grown food; the prevalence of wind farms will continue to grow in the UK.

**6 a** Carbon neutral means adding no carbon dioxide to the atmosphere, overall; if any is added, it is cancelled out by planting trees.

**b** Figure B shows that despite recent growth in renewables, the world still uses a lot of fossil fuels. More than 80% of the energy used in 2016 was from fossil fuels. Other sources of energy would need to rapidly expand for the world to become carbon neutral by 2060. Alternatives to fossil fuels often require expensive infrastructure, so not all countries will be able to afford them. The change may be slow unless mandatory restrictions are put on the use of fossil fuels on a global scale.

# Energy part 2 – Renewables in the UK



## About this unit

In this unit students will find out how electricity is generated from renewable sources in the UK, and how those sources are linked to the UK's location and physical geography.

## Key points

- By 2050 most of the UK's electricity will come from renewable energy sources.
- Six ways that the UK can produce green electricity include: power stations burning biomass or waste; from hydroelectric power stations, wind farms, waves and tides, and solar farms. All except solar farms use turbines to create electricity.
- Highest wind speeds are found in the north and western parts of the UK, and that is where most onshore wind farms are found.
- Most of the UK's solar farms are in the south, east, and central areas where sunlight is strongest.
- 37% of the UK's electricity came from renewable sources in 2019.

## Key vocabulary

green electricity, renewable energy sources, biomass, waste, hydroelectric, wind farm, waves, tide, solar farms, PV cells

## Skills practised in 'Your turn'

- Geographical skills: q1, interpret a diagram; q3, interpret photos; q4, q5, q6, interpret and analyse maps; q6, compare maps; q7, interpret a pie chart
- Numeracy skills: q7, calculate a total percentage
- Literacy skills: q2, define
- Thinking skills: q1, draw a flow chart; q3, q8, make suggestions; q4, q6, describe; q4, q5, q6, q8, explain; q5, q6, make links; q7, give an opinion

## Unit outcomes

By the end of this unit most students should be able to:

- explain the terms given in 'Key vocabulary' above
- draw a flow chart to show how electricity is generated using a turbine
- explain the link between wind speeds and the location of onshore wind farms
- describe and explain the pattern of solar farms in the UK
- give the main source of the UK's electricity in 2019
- list the UK's renewable sources of energy which depend on location and physical geography

## Ideas for a starter

1. Tell the class you are going to do something amazing. Then switch on the light. Ask: What's causing the light? What made the electricity? How did it get here? Can anyone explain?
2. With books closed, show diagram J, showing how electricity is generated using a turbine, on the board. Ask: What else could turn a turbine? A fast flowing river, waves, tides.

## Ideas for plenaries

1. Are biomass and waste *really* renewable sources of energy? Discuss.
2. Play 'Just a minute'. The topic is 'Renewables in the UK'. Students have up to one minute to talk about the topic without hesitation or repetition.

## Further class and homework opportunities

Suggestions 14 and 15 on page 63 of this book

*geog.3 workbook,* page 20

**kerboodle**

### Resources

2.9 Sort it out! – Students match the renewable energy source with its locational requirements.

2.9 Word scramble – Students complete a crossword about renewable energy in the UK.

2.9 Foundation worksheet – Students list types of renewable energy, evaluate the ways we make green energy, and decide on the best location for a solar farm.

2.9 Extension worksheet – Students design a poster promoting renewable energy and the UK's goal to produce most electricity from renewable energy resources by 2050.

### Assessment

2.9 End-of-lesson test – An assignable, auto-marked formative assessment.

## Answers to 'Your turn'

**1** Flow charts will vary, but could include: force such as wind > turbine spins > coil of copper spins > electricity is produced > electricity fed into grid > electricity reaches consumers

**2** **a** Biomass is plant or animal material used as fuel, for example wood, or waste from farms.

**b** A PV cell (photovoltaic cell) converts the energy from sunlight into electricity.

**3** **a** No harmful gases produced by: B, C, D, E, F

**b** No turbines required by: F

**4** **a** The solar farms are more heavily concentrated in the southern half of the United Kingdom; in Wales, there are more on the south coast; in Scotland, there are more on the east coast.

**b** There are more solar farms in the south because the strength of the sun increases as you move closer to the Equator. Scotland is mountainous and cloudier in the west due to prevailing winds blowing moisture in from the sea, so the east has more sunshine.

**5** Yes. There are more onshore wind farms in Scotland because it experiences higher windspeeds. Coastal locations seem to be a favourable site for wind farms for the same reason. There are very few land-based wind farms in the south and east of England, which has lower windspeeds.

**6** Most of the hydroelectric stations are in north-west Scotland. There is also a small cluster in north Wales and a few in Cornwall. This seems to correlate with the location of mountain ranges in the UK, notably: the north-west Highlands, the Grampians, the Cambrian Mountains, and Dartmoor. To explain, students should mention that hydroelectric power is generated when water flows from a reservoir through a dam and spins the turbines. These reservoirs are usually located on high ground in large water catchments. Steep land gradients increase the force of the water.

**7** **a** Gas

**b** 37.1%

**c** Over a third of electricity was produced from renewable sources in 2019, which is a good start. Nuclear is also a relatively low contributor to carbon emissions. The main barrier to becoming carbon neutral is the reliance on gas. The UK would need to quickly develop more renewable energy sites to compensate. Alternatively, it could import more energy which comes from renewable sources. Students might point out that K only relates to electricity but there are other factors which would affect whether the UK could become carbon neutral – such as petrol used in cars and kerosene used by the aviation industry. To become carbon neutral is a challenge and may involve offsetting, eg planting forests.

**8** **a** **i** Solar, Tide, Waves, Biomass. Students could also mention wind and hydro because to a certain extent, these relate to our latitude and the fact we are in a low pressure belt.

**ii** HEP, Wind, Biomass. Biomass is in both categories because it requires a certain climate to grow the crops and low flat land with fertile soil.

**b** Students could choose, for example, a landlocked European country, because that country won't have access to wave or tidal power, or offshore wind farms. However, it may be very well placed for onshore wind farms, or solar power. (The further south, the stronger the sunlight.)

# Energy part 3 – Is solar power the solution?



## About this unit

This unit explains what solar power is, and how sunlight varies around the world. It considers whether solar power is a solution to lack of electricity for poorer countries.

## Key points

- To obtain solar power, energy from sunlight is converted to electrical energy, using special materials contained in PV cells.
- The strength of sunlight varies around the world. Stronger sunlight means solar panels give more electricity.
- Many of the world's poorest countries are in Africa, with strong sunlight.
- Solar power can be used at any scale, so electricity is within the reach of everyone.

## Key vocabulary

solar power, PV cell, solar farm

## Skills practised in 'Your turn'

- Geographical skills: q2, q3, interpret and analyse a choropleth map; q5, q7, interpret photos
- Literacy skills: q3, unscramble a word
- Thinking skills: q1, comprehension; q2, q4, q5, explain; q3, q7, suggest a reason; q4, q7, give an opinion; q4, suggest disadvantages; q5, imagine a scenario; q6, design a vaccine storage unit

## Unit outcomes

By the end of this unit most students should be able to:

- explain the terms given in 'Key vocabulary' above
- describe how solar power is produced
- describe how the strength of sunlight affects the amount of electricity produced in different locations
- describe how the world's poorest countries could benefit most from solar power

## Ideas for a starter

1. With books closed, ask: Tell me what you know about solar power. Include anything you know from outside class. Students could work in pairs or small groups and create a set of statements.
2. With books closed, and without comment, show a set of images of varied uses of solar power. For example: a solar-powered plane, car, toy with waving arms, garden lanterns, fridges. Ask: What theme links all these images?

## Ideas for plenaries

1. What if … solar cells had been developed before the internal combustion engine (which is used in cars and other vehicles)? How would our world be different?
2. Could solar power become the world's main source of energy for the future?
3. When can solar power not be generated? (At night, on dark days, and if sunlight were to be blotted out by a volcanic eruption, for example. But remember it can be stored in a battery, and *used* at night.)

## Further class and homework opportunities

Suggestions 16 and 17 on page 63 of this book

*geog.3 workbook,* page 21

**kerboodle**

### Resources

2.10 Quiz time! – Students complete a quick quiz about solar power.

2.10 Where in the world? – Students complete a quick quiz about solar power around the world.

2.10 Foundation worksheet – Students consider the issues surrounding solar power, and identify the advantages, disadvantages and social impact of solar panels.

2.10 Extension worksheet – Students write a report comprehensively evaluating solar power, giving their own recommendations about what to do in the future.

### Assessment

2.10 End-of-lesson test – An assignable, auto-marked formative assessment.

## Answers to 'Your turn'

**1** A PV cell

**2** **a** Africa

**b** The choropleth map key indicates that solar panels in Africa generate large amounts of electricity, particularly over the Sahara Desert and in the south-west. There is a link between the amount of electricity generated and the strength of sunlight.

**c** Chile, particularly in the northern regions.

**d** Spain is closer to the Equator and so sunlight is stronger than in the UK. This may vary on individual days however, perhaps if there was very thick cloud in Spain but clear skies in the UK.

**3** The jumbled word is 'cloudy'. It is actually cloudy at the Equator (X) due to a low pressure belt which causes air to rise, cool and condense. Y is located further from the sun, but at 30°S there is a high pressure belt and typically clear skies, so the sunlight will be stronger. To aid understanding, diagrams of the global circulation system and pressure belts would help (these are available in Unit 5.2 of the *geog.2* students' book).

**4** **a** Students need to give their own opinions here. Ideas may include: solar power is renewable because sunlight is an infinite resource (compared to fossil fuels, which are finite); there are no emissions associated with solar power; there is no infrastructure required, so PV cells can be used in remote places; solar panels are portable, and do not have to be used in one fixed location; solar technology is getting cheaper.

**b** Solar power is less effective in places which are cloudy. Regions at high latitudes (eg Scandinavia) have some months of the year in complete darkness, so solar power would not work for them in winter. Solar energy cannot be stored easily and it often has to be used straight away.

**5** With solar power, the girls may be able to study in the evenings after dark if the electricity is used to power lamps. They may be able to get mobile phones with Internet access that they could use to further their study and make connections with the wider world.

**6** Students' designs are likely to be portable refrigeration units, powered by PV cells. They would need to be small enough to carry on a motorbike.

**7** Students need to select two of the following options: biomass / waste, HEP, wind, wave, tide, solar. They should then justify why these would be the most effective to turn the world carbon neutral by 2060. Factors to consider could include cost, access to materials, impacts on people and environment.

# Why must we change our ways?



## About this unit

In this unit, students learn more about our impact on other species, how many are in decline, and what we can do to live sustainably.

## Key points

- As population grows, and countries develop, pressure on the Earth's resources grows.
- Scientists warn that a mass extinction is under way – caused by human action.
- Loss of biodiversity is happening in the UK, with 41% of species studied in decline.
- We must learn to live in a sustainable way to protect ourselves and other species.
- One hundred and ninety-three countries have agreed to a set of Sustainable Development Goals to be achieved by 2030.

## Key vocabulary

mass extinction, biodiversity, sustainable, Sustainable Development Goals

## Skills practised in 'Your turn'

- Geographical skills: q3, interpret a photo
- Literacy skills: q1, define; q6, write a speech
- Thinking skills: q2, summarise impacts; q3, make suggestions; q4, write a list; q5, q8, explain; q7, consider personal impacts; q7, give reasons; q9, respond to an opinion

## Unit outcomes

By the end of this unit most students should be able to:

- explain the terms given in 'Key vocabulary' above
- summarise the impact of population growth since 1800, on other species
- list human activities that destroy the habitats of other species
- explain why the loss of species matters
- list some of the things necessary to live in a more sustainable way

## Ideas for a starter

1. With books closed, ask: Who can name some endangered species (Amur leopard, gorillas, giant panda, tigers, snow leopard – named by the WWF as the most endangered)? Why are they in decline?
2. With books closed, write *MASS EXTINCTION UNDERWAY* on the board. What questions would students want to ask to find out more? Write the questions on the board.

## Ideas for plenaries

1. Ask: Do you think we'll ever know how many species there are, or how many are becoming extinct? Explain.
2. Mind movie time. Ask students to close their eyes, and imagine the world 100 years from now where the only plants and animals left are the ones we cultivate and rear for food, and as pets. The rest have gone. The only way we can find out about elephants, lions, and polar bears comes from old films. How does that feel?
3. We could protect species in nature reserves, from which most people are banned. Would you be happy to watch nature on TV, or via webcams?

## Further class and homework opportunities

Suggestions 18 and 19 on page 63 of this book

*geog.3 workbook,* page 22

**kerboodle**

### Resources

2.11 Sort it out! – Students categorise the causes and consequences of the declining bee population.

2.11 Quiz time! – Students match the terms to their definitions.

2.11 Foundation worksheet – Students create a campaign to save species from the sixth mass extinction.

2.11 Extension worksheet – Students write a letter to their MP explaining why we need to protect endangered species, particularly those in the UK.

### Assessment

2.11 End-of-lesson test – An assignable, auto-marked formative assessment.

## Answers to 'Your turn'

1 **a** Habitat – the natural environment of species (its home).

**b** When a large number of species die off over quite a short period of time, for example because of an ice age.

**c** Living sustainably – in such a way that it can be continued into the future without doing harm.

**d** Biodiversity – the variety of living things in a place.

2 As population rises, humans are using more space. Cities expand and more land is used for farming, so we encroach into places that were previously inhabited by other creatures. We also use more of Earth's resources (water, timber, etc). The impact on other species is that many are being driven to extinction, as their habitats are lost. Others are killed or captured by humans. Climate change is also having an impact on many species.

3 **a** Some animals will have died due to the land clearance in photo A. Some will have escaped, but their habitat has been destroyed. As they search for new territory in nearby forests, they will be in direct competition with the animals that already live there, so there will be less food overall.

**b** The trees are likely to have been cut down by humans, for use as timber. Most other plants will have perished in the fire.

4 The list could include: building, mining, farming, fishing, logging, manufacturing, burning fossil fuels.

5 Bees pollinate crops, flowers and trees. The process of pollination fertilises plants and allows them to reproduce. Plants need to be pollinated to produce fruit, vegetables, nuts and other foodstuffs. These are not only for consumption by humans but also for livestock in the meat and dairy industry. (Food security is when people have access, at all times, to enough safe and nutritious food for a healthy and active life.)

6 Students should select one creature and write from its perspective.

7 Unfortunately, most of us are responsible in some way for the destruction of animal habitats. Our settlements are built on land which would once have been wild and inhabited by various creatures. The resources we use (wood, metal, rock, fossil fuels) all require deforestation and / or mining. Land has been cleared to grow the food we eat. The impact of burning fossil fuels is causing climate change and some animals may struggle to survive these changes.

8 Students' answers should focus on the impact to future generations. The term *sustainable* should be used.

9 Answers will vary and should provide a response to one of the statements.

# Using Earth's resources



## Skills practised in check questions

- Geographical skills: q1, q3, interpret a photo; q2, interpret and analyse a scatter graph; q3, interpret a choropleth map
- Literacy skills: q1, define; q5, extended writing
- Thinking skills: q1, make links; q1, q2, q3, explain; q1, q4, suggest reasons; q1, q3, q4, comprehension; q2, suggest an impact; q5, discuss a statement

## Answers for check questions

**1 a i** Food insecurity is when people do not have access to enough safe and nutritious food for a healthy life.

**ii** Locusts can increase food insecurity because they can destroy crops very quickly on a large scale, making food more scarce in the affected region.

**b** *Poor soil* can create food insecurity because crops grow best in nutrient-rich soil. If soil is poor and farmers can't afford fertiliser to enrich the soil, the yields may be poor. *Water stress* can create food insecurity because if there is a lack of water, farmers may not be able to irrigate their land and may struggle to grow healthy food crops. *Poor roads* can create food insecurity if they make it difficult to transport food to market.

**c** As the climate changes, pests may become more common. Locusts breed in warm, wet weather so they may spread to places at higher latitudes and higher altitudes.

**d** One from: importing food; making trade deals to get lower prices; increasing the amount of food grown here.

**e** Hydroponics

**2 a i** People were wealthier in Ghana

**ii** Ghana had a lower % undernourished

**b i** Malaysia

**ii** Less than 5% of the population were undernourished in Malaysia. There are some slightly less wealthy countries where percentages are only 1 or 2%. This hints at some unequal income distribution in Malaysia, with some people below the poverty line in spite of the country's wealth.

**c** Scatter graph

**d** When average wealth per person increases, the percentage of the population which is undernourished decreases. This is a strong negative correlation.

**e** Students should make the link between wealth and food security. Wealthier countries usually have social welfare systems, which take care of the basic needs of their citizens, ensuring they can afford food. With higher average wages, people can afford to buy imported food, so they are not affected by floods, droughts and pests to the same extent as poorer countries. Some students may also point out that the graph shows some residual values (outliers) where wealth is high, but undernourishment is also high. This indicates inequality and a gap between the richest and poorest people in those countries.

**f** One from: undernourished people may not have enough energy for work; children may not be able to focus at school if they are hungry; if undernourishment reaches high levels, a country can suffer famine.

**3 a i** A wind farm generates electricity. It is a 'green' alternative to burning fossil fuels.

**ii** A renewable energy source is one that will not run out.

**iii** Tide and wave power

**b** New coal burning power stations are not permitted in the UK, because the UK is aiming to become carbon neutral. Coal is the 'dirtiest' fossil fuel and burning it produces more $CO_2$ than burning oil or gas. Another reason is that coal is finite and reserves are running out. Wind farms on the other hand, generate energy which is renewable and will help the UK meet its carbon targets.

**c i** Sunlight is renewable because it is a natural source that will not run out. Sunlight is infinite, unlike fossil fuels, which are finite.

**ii** Electricity is obtained from sunlight using PV cells, which contain materials that generate and collect current.

**iii** African countries are 'wealthier' than the UK in terms of strong sunlight because they can generate more energy using solar panels. The amount of energy generated across Africa is variable dependent on climate and cloud cover, but the map shows that solar panels in any part of Africa would generate more electricity than solar panels in any part of the UK.

**4 a** One from: population is rising so we are using more water for agriculture, industry and domestic use; global wealth levels are rising which also increases water usage; water sources may be polluted by industry; aquifers are being depleted faster than they are being replenished; climate change is causing more droughts in some places; water sources are not spread evenly so some places experience more stress than others.

**b** The main use for fresh water is agriculture.

**c i** C is a photo of the ocean. Sea water is salty, not fresh.

**ii** Desalination

**5** For this extension question, students are asked to discuss a statement. Good answers will consider different angles to the argument and cover a range of issues.

Answers might mention: As population has risen and new technology has created a higher demand for resources, our planet is being put under great stress. To reduce this stress, people can try to use less water, less electricity, and less packaging; co-operate more over issues like climate change and access to resources; protect other species and vital biodiversity.

# Further suggestions for class and homework

Many of these suggestions are addressed to your students. Where research or further resources are needed, the Internet will almost certainly provide the answer.

The suggestions are graded *, **, *** according to level of difficulty. Many are suitable for all levels, and can be differentiated by outcome.

## Earth's natural resources

1 **Alphabet run** A challenge. Design an illustrated alphabet run for Earth's natural resources, which we make use of directly or indirectly. So Z could be zinc ore (from which we get zinc) or zooplankton (marine organisms which fish feed on, and whose remains turned into oil over millions of years). **/***

2 **Your global footprint** You make use of natural resources in other countries too. Try to find out where the food items you eat came from, over say two weeks. Make a note of those which were imported, and from where. You could take photos too. Soil, sunlight, air, and water in other countries were used to grow them. */**/***

## Water part 1 – Do we have enough?

3 **Compare** Compare countries on Google Earth to see how liquid fresh water is unevenly distributed. For example Chad with Bangladesh or Canada. Look at Greenland too. What can you say about the fresh water there? **/***

4 **Show another way** Use squared paper to show the percentages of Earth's water that are salty, and fresh. Use a different colour for each. Then add fine dots and lines to show fresh water that is frozen, and below ground. */**/***

## Water part 2 – Emptying the Ogallala?

5 **Brad's story** Imagine it's 25 years into the future. You are Brad, Doug's son from page 28 of the students' book. Doug died a long time ago. How is your life? How is your farm? How is the Ogallala? What has changed? **/***

6 **The Grapes of Wrath** This novel by John Steinbeck is set partly in Oklahoma (an Ogallala state) during the Dust Bowl era. Find a suitable extract to read to the class. */**/***

## Water part 3 – Tackling water stress

7 **Desalination** How does it work? Prepare a very simple explanation, for a ten-year-old. Then give the pros and cons of obtaining fresh water this way. ***

8 **Water wars** Could there be wars over fresh water one day? With the help of an atlas, suggest two or three sets of countries which may already be storing up tensions between them over water use. What criteria did you use for choosing them? ***

## Food part 1 – Feeding the UK

9 **Crossword** Make up a crossword for all the terms you've met in this unit. Give it to a partner to try. */**/***

## Food part 2 – Food insecurity, a global issue

10 **More on Malawi** Malawi is where Chifundo comes from. You have to write about it, for a *geog.1* textbook for Year 7. First, brainstorm which aspects to cover. Then write a two-page spread about it, in geog. style. Add maps and photos. (You can use this in preparation for work in chapter 4.) **/***

## Food part 3 – Tackling food insecurity

11 **Podcast** Prepare a 2-minute audio clip for a podcast on ways to tackle food insecurity. You want to make people sit up and listen! Record your clip. **/***

## Energy part 1 – Goodbye to the fossil fuels?

12 **The motor car** The invention of the motor car helped the oil industry take off. Create an illustrated history of the motor car, as a long wall frieze with pictures and notes. */**/***

13 **The UK's oil deposits.** The UK has some oil. Where? And how much? Mark the oil fields on a map. What gives the UK the right to these deposits (which lie under the sea)? ***

## Energy part 2 – Renewables in the UK

14 **Renewable energy near you** Is renewable energy being exploited near you? Is there a wind or solar farm, or a hydroelectricity station? Does your home have solar panels? Find out. Take a photo if you can. Draw a sketch map of the location and explain why it is (or isn't) appropriate. **/***

15 **Video** Prepare a video for a 2-minute TV slot, to encourage the UK to switch fast to renewable energy sources. */**/***

## Energy part 3 – Is solar power the solution?

16 **Inventor** If you didn't do question 6 on page 43 of the students' book … design an exciting new product powered by solar energy. Draw a diagram of it, and write a press release explaining how it will be used. A prize for the best invention? */**/***

17 **Power in the future** It's the year 2200. Where are we getting our energy from? Use your imagination to describe the types of transport people use, the houses they live in, how they cook, and keep their homes warm (or cool). **/***

## Why must we change our ways?

18 **Poster** You plan to save an endangered species. Make up a name and slogan for your campaign. Create a poster giving five facts about your chosen species, how you plan to save it, and how other people can help you do this. */**/***

19 **The bee** We'd be in big trouble if bees disappeared. Find out about them, how and where they live, why they are so important, and why are they in decline. Decide who your target audience is, then present this information to suit. ***

# 3 Earning a living

## About this chapter

### Addressing the KS3 Programme of Study

- Understand, through the use of detailed place-based exemplars at a variety of scales ... human geography relating to economic activity in the primary, secondary, tertiary, and quaternary sectors; and the use of natural resources

### About the content

- We have chosen Doncaster as an example of a British town which underwent big changes, with the rise and decline of manufacturing. Today its location is playing a critical part in its regeneration.
- This chapter investigates globalisation, the impact of COVID-19 on the high street and employment, and what jobs will look like in the future.

## The big picture

These are the key ideas in this chapter.

- Jobs are grouped into employments sectors: primary (eg farming), secondary (eg construction), tertiary (services), and quaternary (often dependent on R&D).
- In employment structure pie charts, quaternary is treated as a subset of tertiary for a number of reasons, including the difficulty of obtaining separate statistics for the quaternary sector.
- Changes to employment structure, and future employment options, are driven by technology, globalisation, and the action of governments.
- Globalisation is pushed by businesses and governments, and there are gains and losses as a result.
- COVID-19 was an example of globalisation which affected employment, both in the UK and around the world.

## The chapter outline

Use this to give students a mental roadmap for the chapter.

3 **Earning a living** The chapter-opening photo in the students' book is an important resource. See opposite for using it.

3.1 **What kind of work do we do, in the UK?** Looking at primary, secondary, tertiary, and quaternary sectors of employment

3.2 **The UK's changing pattern of employment** How it has changed from 1600 to today

3.3 **Example: change in and around Doncaster** The changes in employment experienced by Doncaster

3.4 **Why do employment patterns change?** The main drivers of change – technology, globalisation, and governments

3.5 **A closer look at globalisation and the UK** Who pushes it, and will it continue?

3.6 **Is your mobile linked to globalisation?** Workers in all employments sectors and in many countries play a part in bringing us our mobile phones

3.7 **Employment and the shock of COVID-19** How the pandemic affected the high street and employment

3.8 **So what kind of work will *you* do?** The drivers of change, and possible future jobs

3 **Earning a living check questions** A page of questions to check knowledge, understanding and skills across the chapter, with an opportunity for extended writing

## Objectives and outcomes for this chapter

| Objectives | Unit | Outcomes |
| --- | --- | --- |
| Most students will understand: | | Most students will be able to: |
| • that jobs can be grouped into different sectors | 3.1 | • name the four employment sectors; identify employment sectors for a variety of jobs |
| • that employment structure changes over time | 3.2 | • outline how, and why, employment structure has changed in the UK, from 1600 to today; describe the employment structure for the UK today |
| • how and why employment changed in British towns and cities (Doncaster as an example) | 3.3 | • explain how the loss of industry leads to decline and deprivation; describe how Doncaster's location has brought new jobs |
| • what drives changes in employment patterns | 3.4 | • define technology and globalisation; explain how the three main drivers cause changes in employment patterns |
| • who pushes globalisation and whether it can be slowed down | 3.5 | • suggest how globalisation benefits the UK; who loses; and how globalisation could be slowed down |
| • how a mobile phone is linked to globalisation | 3.6 | • describe how people in all employment sectors are involved in bringing us mobiles; and that workers in many countries are involved in their production |
| • how a pandemic affected employment | 3.7 | • explain why the COVID-19 pandemic has been linked to the globalisation of air travel; and the impact it had on shopping and the high street |
| • what work might be like in the future | 3.8 | • describe how technology, globalisation and the government can drive changes in jobs; give an example of artificial intelligence; define the gig economy and green jobs and give examples of each |

These tie in with the outcomes for each unit in this teacher's handbook.

## Assessment for this chapter

There are assessment materials for this chapter on *geog.3 Kerboodle Lessons, Resources, & Assessment*. They include interactive auto-marked end-of-lesson and chapter-level tests, a printable exam-style end-of-chapter paper (with a mark scheme), and a printable self-assessment checklist. See also the notes on pages 14–17 of this book.

## Getting ready for this chapter

There is a scheme of work for this chapter, plus a glossary and glossary worksheet, on *geog.3 Kerboodle Lessons, Resources, & Assessment*.

## Using the chapter-opening photo

Ask students:

- What is this person doing? Does it count as work?
- About how many people in the UK work for a living? Guess!
- How would it affect the UK if everyone stopped working?

The photo shows a street entertainer dressed as Yoda from *Star Wars*, outside the National Gallery in Trafalgar Square, London.

# What kind of work do we do, in the UK?



## About this unit

This unit looks at the type of work that people do, where they work, and what the economy is.

## Key points

- People in work provide us with goods and services; and the income tax they pay helps the government to run the country.
- Jobs can be grouped into four employment sectors – primary, secondary, tertiary, and quaternary.
- The sector that people work in can determine where they work. Some jobs, such as those in the primary sector, have to be done in particular locations; others, eg some in the tertiary sector, can be done from any location.
- The economy comprises all the activities that occur in a country in producing, buying, selling and distributing goods and services.

## Key vocabulary

goods, services, income tax, primary, secondary, tertiary, and quaternary employment sectors, raw materials, research and development (R&D), science parks, economy

## Skills practised in 'Your turn'

- Literacy skills: q1, q6, define
- Thinking skills: q1, list examples; q2, comprehension; q3, q4, classify; q4, suggest examples; q5, suggest factors; q6, decide and explain

## Unit outcomes

By the end of this unit most students should be able to:

- explain the terms given in 'Key vocabulary' above
- name the four employment sectors
- identify jobs in different employment sectors
- give four factors which influence where jobs are located
- explain what the economy is

## Ideas for a starter

1. Ask: What jobs are needed to get the bread for your morning toast into your home? Begin with the ingredients – flour, yeast, salt, oil, etc.
2. Ask: Can you name any jobs which are unlikely to be found in this area? Ask students to explain their choices. Keep asking 'Why?' until students drill down to the key factors.

## Ideas for plenaries

1. Imagine the 33 million people at work in the UK. Why are all those workers good for the economy? What would happen if they all lost their jobs?
2. Thinking ahead. Ask students what percentage of people they think work in each employment sector in the UK today? Why do they think this?

## Further class and homework opportunities

Suggestions 1–4 on page 83 of this book

*geog.3 workbook*, page 23

**kerboodle**

### Resources

3.1 Word scramble – Students complete a crossword of key terms about the UK economy.

3.1 Sort it out! – Students sort jobs into their employment sectors.

3.1 Foundation worksheet – Students draw pictures depicting jobs that would be found in each of the four job sectors.

3.1 Extension worksheet – Students conduct a survey to find out what jobs their classmates' parents have, depict the results in a graph, and write a paragraph interpreting their findings.

### Assessment

3.1 End-of-lesson test – An assignable, auto-marked formative assessment.

## Answers to 'Your turn'

**1** **a** **i** Goods are physical objects, such as pens and cars, that are bought and sold.

**ii** Services are activities carried out to meet needs, such as teaching people, or helping sick people get better.

**b** Four examples of goods: Clothes, electronics, cosmetics, furniture

Four examples of services: Driving, cleaning, counselling, security

**2** **a** Primary, secondary, tertiary, quaternary

**b** The secondary sector includes manufacturing

**3** **a** Quarry worker – primary

**b** Person developing a new camera – quaternary

**c** Postal delivery worker – tertiary

**d** Carpenter – secondary

**e** Lighting director – tertiary

**f** Hairdresser – tertiary

**4** Answers will vary. Encourage students not to repeat previous examples. Two for each part, plus the sector.

**a** For example – nail technician – tertiary. This may be an interesting discussion considering how things changed during the pandemic. Many services we previously expected to experience face-to-face were moved online, such as doctor's appointments, guitar lessons, etc.

**b** For example – quantity surveyor – tertiary

**c** For example – vet working in a surgery – tertiary

**d** For example – software developer – quaternary

**5** Students should make suggestions such as: close to the source of raw materials or manufactured goods required in the process; close to a good road network for transporting the cars to market; close to an urban area for access to workforce.

**6** **a** This term is used for the whole network of activities involved in producing, supplying, and consuming goods and services, during which money changes hands.

**b** **i** Yes, the teacher is being paid to educate people, which is a service.

**ii** Yes, the footballer is being paid to entertain people, which is a service.

# The UK's changing pattern of employment

Help at a glance

## About this unit

This unit describes how the UK's employment structure has changed since 1600 (when a large % of people were in farming). It includes a graph showing the pattern of change.

## Key points

- Employment structure changes over time – driven by new inventions, for example.
- It is estimated that in Britain in 1600 around 70% of the working population was in the primary sector – mainly in farming.
- By 1850, the % in the primary sector had shrunk because of changes in farming. And the secondary sector had grown, thanks to the Industrial Revolution.
- By 1970, manufacturing was in decline, thanks largely to competition from abroad.
- Today, about 83.6% of the workforce is in the tertiary / quaternary sector, 15.1% in the secondary sector, and just 1.3% in primary.

## Key vocabulary

primary sector, secondary sector, tertiary sector, quaternary sector, Industrial Revolution, employment structure, model graph, Clark-Fisher model

## Skills practised in 'Your turn'

- Geographical skills: q3, interpret and analyse a model
- Thinking skills: q1, give examples; q2, suggest reasons, suggest consequences; q3, explain; q4, make suggestions; q5, discuss an idea

## Unit outcomes

By the end of this unit most students should be able to:

- explain the terms given in 'Key vocabulary' above
- outline how and why the UK's employment structure changed, from 1600 to today
- sketch a rough line graph to show the pattern of change
- say that employment structure has been changing in most other countries too, but not necessarily in the same way or at the same rate as in the UK

## Ideas for a starter

1. With books closed, ask: What do you think was the most common job in Britain 400 years ago? How many jobs can you think of, which did not exist 400 years ago? Accept all answers and write them on the board. You can come back to them later.
2. With books closed, write the title of today's unit on the board, without vowels. What is it? *Th K's chngng pttrn f mplymnt*

## Ideas for plenaries

1. If you chose starter 1 you could follow it up now.
2. Look at the graph on page 53. How do you think it will look 100 years from now? You could ask a student to sketch it on the board, and extrapolate into the future.
3. *We will always need plumbers and electricians and farmers.* Discuss.
4. Summarise the key points of today's lesson; in five sentences; then in five words; then in one word!

## Further class and homework opportunities

Suggestions 5–7 on page 83 of this book

*geog.3 workbook*, page 24

**kerboodle**

### Resources

3.2 Skills workshop – Students interpret a graph of the UK employment structure.

3.2 Quiz time! – Students complete the explanation of the UK's changing employment structure.

3.2 Foundation worksheet – Students complete a Venn diagram to show the differences and similarities between the types of jobs in 1850 and 2015.

3.2 Extension worksheet – Students write diary entries describing life as a 13-year-old in 1600, 1850, and today.

### Assessment

3.2 End-of-lesson test – An assignable, auto-marked formative assessment.

## Answers to 'Your turn'

**1 a** Any three goods not yet invented by 1850. For example: microwave, mobile phone, electric toothbrush.

**b** Any three services which did not exist in 1850. For example: pilot, broadband technician, DJ.

**2 a** It means approximately.

**b** 1.3%

**c** Two from: mining, fishing, forestry, oil extraction

**d** It has fallen greatly from 1600

**e** Students could choose from among all these reasons (but not necessarily in this detail):

*Farming:* it has become mechanised and much more efficient; crop strains have been improved; artificial fertilisers are widely used; so the same amount of food can be produced by far fewer people. At the same time more food is imported, and supermarkets are pushing down the prices paid to farmers. So we don't need so many farmers, and people are less attracted to farming because it's not so profitable.

*Mining:* there used to be over 1 million people working in coal mining in the UK, but now almost all the mines are closed.

*Fishing:* industrial fishing has also become more mechanised, needing fewer people.

**f** Students should suggest three impacts. If the primary sector shrank to zero in the UK: people working in this sector would lose their jobs; training opportunities would need to be provided to help them learn new skills; agricultural land would be abandoned or could be rewilded; the UK would need to import all food, energy and resources; importing reduces food security.

**3 a** the % of the workforce employed in the secondary sector

**b** The line rises to a maximum around 1900, then falls again.

**c** The % rose as a result of the Industrial Revolution – many factories opened, and Britain exported factory goods everywhere. But then the % began to fall as other countries started to manufacture their own goods, and export them cheaply. British factories could not compete. Many closed.

**d** The blue line could rise if the secondary sector grew again in the UK. This would happen if the UK opened more factories or if the construction sector expanded.

**4** Any two jobs associated with the Internet, eg website designer, website author, website programmer, online shopping-site employee

**5** Some students might think it's a very good idea. Ask what they would do instead.

# Example: change in and around Doncaster

Help at a glance

## About this unit

This unit looks at Doncaster – a town which has experienced big changes in employment.

## Key points

- Beginning as a Roman fort on the route from London to York, Doncaster developed as a market town. With the Industrial Revolution, it became an industrial and mining centre.
- Around 1970 decline set in, in both manufacturing and mining.
- People lost their jobs, leading to poverty, ill-health and deprivation.
- Now Doncaster is regenerating itself. Thanks to its location and excellent transport links, it has become a logistics hub for storing and distributing goods. It still has some manufacturing, but in new fields.

## Key vocabulary

industrialisation, industrial decline, logistics hub, i-Port, innovation, unemployment, wages

## Skills practised in 'Your turn'

- Geographical skills: q3, interpret a map
- Literacy skills: q2, define
- Thinking skills: q1, classify; q2, q3, q4, explain

## Unit outcomes

By the end of this unit most students should be able to:

- explain the terms given in 'Key vocabulary' above
- name two activities that developed in Doncaster as a result of the Industrial Revolution
- explain how the loss of industry can lead to deprivation
- explain why companies are attracted to Doncaster as a warehouse centre

## Ideas for a starter

1. With books closed, hand out atlases. Give students five minutes to come up with as many geographical facts as they can about Doncaster – at least four. (In England, on the River Don, on a motorway, on lowish land, about x km from the coast, near xxxxx and so on.)
2. With books closed, show a set of images of Doncaster, including aerial images – without naming the town. For example: the old railway works, horse racing, the i-Port, and so on. Ask: What is your impression of this town? (Students could build up a spider map.)

## Ideas for plenaries

1. Spend time on the map. Ask questions such as: name five cities to which Doncaster is linked by motorway; name the ports to which it is linked; why is Doncaster better located than say Leicester or Cardiff, as a warehousing hub?
2. Explore Doncaster on Google Earth. Look for transport links, the River Don, signs of warehousing, sport and recreation, and farming outside the town. Find Doncaster Sheffield airport, and villages and small towns around Doncaster which once had mining … for example Mexborough, Thorne, Rossington, and Hatfield (the last colliery to close).

## Further class and homework opportunities

Suggestions 8–10 on page 83 of this book

*geog.3 workbook*, page 25

**kerboodle**

### Resources

3.3 Sort it out! – Students label flow diagrams to explain industrial change.

3.3 Case study: Doncaster – Students describe how Doncaster has changed over time.

3.3 Foundation worksheet – Students match figures with statements about Doncaster and give reasons why the city might be a good place to locate a business.

3.3 Extension worksheet – Students expand on the history of Doncaster, give reasons why it might be a good place to locate a business, and identify problems the city still faces.

### Assessment

3.3 End-of-lesson test – An assignable, auto-marked formative assessment.

## Answers to 'Your turn'

**1 a** Building steam trains (secondary sector) and coal mining (primary sector)

**b** Tertiary sector

**2 a** The state of looking for work and not being able to find it

**b** An area of deprivation is one where people suffer disadvantages such as high unemployment, poor health, and poor housing.

**c** Loss of industry can lead to deprivation because: when industry closes down, people lose their jobs and may not be able to find work in other sectors if they don't have the appropriate skills; unemployed people may become depressed which can affect their health in a variety of ways; people will have less money and they might have to cut back on essentials such as food and heating; local shops and services will suffer if people can't afford to use them; local councils will have less income if businesses start closing; some people may move elsewhere to look for work; derelict buildings might become common; crime levels may rise; a place may experience a 'spiral of decline'.

**3** Students should mention that Doncaster is in quite a central position in the map area, well connected by transport links to the UK's biggest cities. The motorway is vital for delivering goods to company stores. IKEA and most other companies get goods made in other countries and shipped in, so excellent links to ports are essential. Doncaster also has unemployed people ready to work.

**4 a** The companies pay staff more, so staff spend more, so local shops and other services benefit. New shops and services may open (eg gyms, restaurants). Well-paid jobs will attract skilled people to Doncaster. Successful businesses inspire others to start up too, so the effect spreads. The amount the local council earns from businesses increases, so it has more to spend.

**b i** It will attract students, who will need food, flats to rent, and so on. It may also attract companies keen to link up with the university. And universities often start commercial businesses, spun off from their research.

**ii** IT companies usually need skilled people and pay well. They often become successful, so office buildings with low rents to attract IT companies could be a good move.

# Why do employment patterns change?



## About this unit

This unit examines the main reasons why employment patterns change.

## Key points

- There are three main reasons why employment patterns change, to do with technology, globalisation, and governments.
- Technology has led to falls in the percentage of people employed in the primary and secondary sectors, but rises and growth in the tertiary and quaternary sectors.
- Globalisation and global trade initially led to a rise in UK manufacturing, but then led to its decline.
- Governments can influence employment patterns by offering finance and grants.

## Key vocabulary

technology, globalisation, de-industrialisation, aerospace, pharmaceuticals, governments

## Skills practised in 'Your turn'

- Literacy skills: q1, q2, define
- Thinking skills: q1, give examples; q2, decide, then justify; q3, q4, give reasons; q4, explain; q5, provide an opinion, give reasons

## Unit outcomes

By the end of this unit most students should be able to:

- explain the terms given in 'Key vocabulary' above
- give examples of how technology has affected two different employment sectors
- explain how globalisation has helped to reduce the percentage of workers in manufacturing in the UK in recent decades
- explain how governments can influence employment patterns

## Ideas for a starter

1. Show students items such as a cup, a T-shirt, a saucepan – none of them made in the UK. For each item ask: Where do you think this was made? Point out that all of these items are easy to make, but they were all imported. Why?
2. Ask students to define the term 'technology'. Then ask: Who can suggest how technology has affected employment in the primary and secondary sectors over the last 200, or so, years.

## Ideas for plenaries

1. People are paid more in the UK than in many other countries. But some savings could be made by having more factories here instead of importing goods. What kind of savings?

   (No shipping costs to import goods, or import taxes. There would be more jobs, so the government might have to make fewer benefit payments, plus income tax might increase.)
2. Use 'Your turn' question 4 as the basis for a class discussion.

## Further class and homework opportunities

Suggestions 11–13 on page 83 of this book

*geog.3 workbook*, page 26

**kerboodle**

### Resources

3.4 Sort it out! – Students complete a paragraph about globalisation and employment in the UK.

3.4 Quiz time! – Students complete a quick quiz about employment patterns.

3.4 Foundation worksheet – Students define key terms from pages 54 and 55 of *geog.3* and complete a paragraph on how globalisation has affected employment in manufacturing.

3.4 Extension worksheet – Students write four short speeches that argue for and against the idea that technology and global trade has been good for UK employment.

### Assessment

3.4 End-of-lesson test – An assignable, auto-marked formative assessment.

## Answers to 'Your turn'

**1 a** Technology is where we put scientific discoveries and inventions to practical use.

**b i** Two-hundred and fifty years ago it was the late 1700s. New technology at the time was being developed in the manufacturing industry and catalysed the Industrial Revolution. Examples include: the flying shuttle, the spinning jenny, and the spinning mule.

**ii** The computer, the Internet and the smartphone are all good examples.

**2 a** The increasing movement of goods, services, people, money, and ideas around the world, making it more connected.

**b i** Holiday abroad – yes, because it involves a flow of people and money. It is due to advancements in transport that we can travel more easily to distant places.

**ii** Using YouTube – yes, because it involves a flow of information. It is due to advancements in technology that we can watch videos posted by people from around the world.

**iii** Buying a T-shirt made in Bangladesh – yes, because it involves a flow of goods and money. We are interconnected to other countries through flows that relate to trade.

**iv** Buying eggs from a local farm – no, unless the chickens are fed imported food. There are no global flows in this transaction, the product is locally produced.

**3** Fewer people now work in manufacturing in the UK due to the process of de-industrialisation. This took place over several decades. Many factories closed down due to competition from overseas industry. Another reason is the use of robots in modern manufacturing, for example, robots can be used for welding in car assembly plants.

**4** British chain stores import clothing from Bangladesh because workers in Bangladesh's textile factories are paid much less to do the same job as a British worker would be paid. This means that the costs of making the garments is cheaper overall and this offsets the cost of shipping. Most consumers want to pay less, so imported clothing is more competitive.

**5** Students might think more factories are a good idea. They provide employment; we can make goods instead of importing them; and we can export goods for profit. Others might suggest that the goods produced in British factories would cost more, so consumers would not want to buy them.

# A closer look at globalisation and the UK



## About this unit

This unit looks at globalisation – how it has sped up, and why, and its impact on the UK.

## Key points

- The UK is connected to the rest of the world as a result of the process of globalisation.
- The pace of globalisation has increased rapidly over the last 60–70 years as a result of containerisation, information technology (IT), and air travel.
- Globalisation is encouraged by the actions of big businesses (eg TNCs) by increasing profits, and governments if it helps their economies.
- Individuals and countries can both win and lose as a result of globalisation.
- Globalisation can be slowed down, or curbed as a result of certain events, such as the COVID-19 pandemic, and by the actions of governments.

## Key vocabulary

globalisation, containerisation, information technology (IT), transnational corporations (TNCs), Brexit, tariffs, trade wars

## Skills practised in 'Your turn'

- Geographical skills: q6, interpret an image
- Literacy skills: q6, justify the extent
- Thinking skills: q1, q4, q5, explain; q2, comprehension; q2, list examples; q3, match comments to criteria; q4, suggest examples; q6, provide an opinion

## Unit outcomes

By the end of this unit most students should be able to:

- explain the terms given in 'Key vocabulary' above
- explain how containerisation has helped to speed up the rate of globalisation
- give five examples of TNcs
- give two benefits and two disadvantages of globalisation
- explain how events or government actions can slow down globalisation

## Ideas for a starter

1. Show students a range of logos, eg Nike, Apple, McDonalds, Amazon, etc. and ask them to match them to the well-known brands. Ask: What do these companies have in common? (They are TNCs.)
2. Use 'Your turn' question 4 as a starter activity.

## Ideas for plenaries

1. Ask students to write 'Globalisation and the UK' in the middle of their page. Give them two minutes to create a mind map.
2. Show the image of the globe nestled in a purse from the top of page 56 in the students' book on the whiteboard. Then do 'Your turn' question 6.

## Further class and homework opportunities

Suggestions 14 and 15 on page 83 of this book

*geog.3 workbook*, page 27

kerboodle

### Resources

3.5 Sort it out! – Students sort the economic gains and losses caused by globalisation.

3.5 Quiz time! –Students complete a quick quiz about TNCs.

3.5 Foundation worksheet – Students write and draw the ways in which globalisation has affected their day.

3.5 Extension worksheet – Students explain the positive and negative consequences of globalisation in the UK.

### Assessment

3.5 End-of-lesson test – An assignable, auto-marked formative assessment.

## Answers to 'Your turn'

**1** Containerisation has helped to expand trade by making the process of transporting goods much easier and cheaper. The containers are a standard shape and size so they can be moved on and off ships, trains, and lorries at ports all over the world, if the appropriate machinery has been installed. Goods can be stacked up inside the containers, which hold an enormous amount.

**2 a** Transnational corporation

**b** All TNCs have operations (factories, shops, offices) in more than one country.

**c** Five examples of TNCs, ideally from different sectors. For example: Apple, Disney, KFC, Toyota, Shell.

**3 a** Any two that show benefits of globalisation such as:

'My company gets its accounts done in India. Much cheaper than here.'

'The UK has earned a great deal from overseas tourists.'

**b** Any two that show harmful impacts of globalisation such as:

'My business failed because a foreign TNC set up here. It took all our customers.

'I lost my job because my company could not compete with cheap imports.'

**c** Any two that show concerns about TNCs such as:

'Many of the things we import are essential. What if we can't get them? That worries me.'

'I think these big TNCs are far too powerful.'

**4 a** Amazon benefits the UK by employing workers in its warehouses and distribution centres; by providing cheap goods for UK consumers; and by providing access to global products.

**b** Amazon also has negative impacts which are mostly associated with the challenges for small UK businesses that can't compete with its low prices, its marketing, its fast shipping and its convenient website. This causes UK businesses to struggle or close down. Students may also mention climate change associated with shipping goods from faraway places.

**5** High tariffs could help slow globalisation by putting taxes on foreign products so that they cost more in shops. This would mean people are more likely to buy local products, as they will be cheaper.

**6** The image of the coin purse represents the 'shrinking world' concept – globalisation has made faraway places seem closer together, and our planet feel smaller. Students must state and justify the extent to which the image represents globalisation. Some may say it sums up globalisation to a great extent because the depiction of a small Earth shows that people now have access to information, goods, and services from all over the world in very short time scales. Others may say it sums up globalisation only to a limited extent because the image shows a purse which emphasises money more than any other elements of globalisation such as people, goods, and information.

# Is your mobile linked to globalisation?



## About this unit

This unit shows how people in every employment sector, and in different countries, play a part in bringing us mobile phones.

## Key points

- The design and development of mobile phones, apps, functions and networks, take place in the quaternary sector.
- Workers in the primary sector extract the many raw materials from which mobile phones are made.
- Workers in the secondary sector make and assemble the phone components.
- Workers in the tertiary sector sell phones and call plans, provide support, and so on.
- Apple designs and markets iPhones – but does not make them. It arranges component manufacture and assembly in factories around the world. And sells goods and services everywhere – an example of globalisation.

## Key vocabulary

primary sector, secondary sector, tertiary sector, quaternary sector, rare earths, globalisation

## Skills practised in 'Your turn'

- Geographical skills: q2, q3, interpret a map
- Literacy skills: q1, classify
- Thinking skills: q1, q2, q3, q5, explain; q2, q4, suggest reasons; q3, deduce

## Unit outcomes

By the end of this unit most students should be able to:

- explain the terms given in 'Key vocabulary' above
- describe how all four employment sectors are involved in bringing us a mobile phone
- name six countries where Apple components are made
- decide and explain whether or not Apple is a manufacturing company
- explain why Apple's business is an example of globalisation

## Ideas for a starter

1. Hold up an iPhone if you have one, or show an image of one. Ask: Who designed it? Who markets it? Who made it? Apple does not actually make its phones.
   The components are made and the phones are assembled in other people's factories.
2. With books closed, hold up a mobile. Ask: What kinds of workers help to bring mobiles to us? Make notes on the board.

## Ideas for plenaries

1. Use question 2 of 'Your turn' as a plenary. Then ask more questions about the map. For example: Which country with over 1 billion people had no Apple store in 2020? (India)
2. Is the Apple map for 2030 likely to be different from the one for 2020? In what way?
3. Why does Apple use other people's factories to make its mobiles?

## Further class and homework opportunities

Suggestions 16–18 on page 83 of this book

*geog.3 workbook*, page 28

kerboodle

### Resources

3.6 Quiz time! – Students complete a quick quiz about mobile phone production.

3.6 Where in the world? – Students complete a quick quiz about global Apple production.

3.6 Foundation worksheet – Students chart how employment sectors help to produce the iPhone and examine how the iPhone and smartphone apps contribute to globalisation.

3.6 Extension worksheet – Students examine the link between smartphones and globalisation.

### Assessment

3.6 End-of-lesson test – An assignable, auto-marked formative assessment.

## Answers to 'Your turn'

**1 a** Workers in the primary sector contribute by extracting raw materials that are used to make smartphones. These include oil, which is used to make plastic; rock, which contains minerals that make glass; and metals.

**b i** Secondary

**ii** Tertiary

**2 a** The USA. Apple is an American company, with its headquarters in the USA. The USA is the third most populous country in the world, and people are well-off, on average, so they can afford Apple products.

**b** UK

**c** Africa; fewer people are likely to be able to afford Apple products.

**3 a** Any six from: Austria, Belgium, Brazil, Cambodia, China, Costa Rica, Czech Republic, France, Germany, Hungary, Indonesia, Ireland, Israel, Italy, Japan, Malaysia, Malta, Mexico, Morocco, the Netherlands, Norway, Philippines, Poland, Portugal, Puerto Rico, South Korea, Singapore, Taiwan, Thailand, UK, USA, Vietnam.

**b** Students should deduce that wages are lower in China than in the USA. This is the main reason why many Apple products are assembled in China.

**4** Some American politicians may prefer Apple products to be made in the USA because this would provide jobs for US workers. Moreover, the USA and China are both global superpowers, so the USA may prefer not to contribute to China's economic growth, in case it becomes more powerful than the USA.

**5** False. Apple arranges manufacturing in many countries via global networks but it does not manufacture the products itself; at Apple HQ, engineers research and develop new technology. It would be more accurate to say that Apple is a technology company.

# Employment and the shock of COVID-19



## About this unit

This unit explores how the COVID-19 pandemic – an example of globalisation – affected employment in the UK.

## Key points

- The COVID-19 pandemic killed over 127 000 people in the UK by May 2021, and caused hundreds of thousands of people to lose their jobs.
- The high street was already changing when the pandemic hit. COVID-19 and lockdowns caused more shop closures.
- While jobs on the high street were lost, others were created as a result of increased online shopping.
- Experts predict that as high streets recover, they will become a mix of shops, coffee bars, restaurants, apartments, and leisure facilities.
- COVID-19 proved that with the help of technology, people could effectively work from home. In future it's likely that working from home will become the norm – for at least part of the time.

## Key vocabulary

pandemic, high street, chain store

## Skills practised in 'Your turn'

- Geographical skills: q2, use local place knowledge
- Literacy skills: q1, q2, q3, define
- Thinking skills: q1, consider a link; q4, make a comparison; q5, imagine a scene and sketch it; q6, suggest advantages and disadvantages; q7, decide, give reasons; q8, give an opinion, discuss

## Unit outcomes

By the end of this unit most students should be able to:

- explain the terms given in 'Key vocabulary' above
- describe how the high street had been changing before COVID-19
- give examples of some of the jobs created as a result of the move to online shopping
- give examples of uses for empty shops that would help to attract different groups of people back to the high street
- give two advantages, and two disadvantages, of working from home for an office worker

## Ideas for a starter

1. Ask students to describe the type of businesses on your local / nearest high street. How have they changed over the last few years?
2. Show students a selection of the images from the material on 'How COVID-19 affected the high street.' Ask them what links these images?

## Ideas for plenaries

1. Design your own high street! Ask students to design a high street with all the shops or facilities they would like to see there. They could write about it, draw it, create a poster with some found images, etc.
2. Working from home – for or against? Hold a class discussion.

## Further class and homework opportunities

Suggestions 19 and 20 on page 83 of this book

*geog.3 workbook,* page 29

**kerboodle**

### Resources

3.7 Word scramble – Students complete a crossword of key terms about UK employment and COVID-19.

3.7 Sort it out! – Students sort the advantages and disadvantages of studying from home.

3.7 Foundation worksheet – Students use a newspaper article template to write about how the COVID-19 pandemic has affected the UK economy.

3.7 Extension worksheet – Students write a newspaper article explaining the impact of the COVID-19 pandemic on the UK economy.

### Assessment

3.7 End-of-lesson test – An assignable, auto-marked formative assessment.

## Answers to 'Your turn'

**1 a** A pandemic is a disease that spreads over a very wide area, or the whole world, and affects a large number of people.

**b** The COVID-19 pandemic has been linked to air travel because people travelled from places where the virus was spreading, and took it with them to other places. It spread quickly to other people, who may also have taken flights without knowing they were transporting the virus.

**2 a** 'The high street' means the heart of a town or city, where the main shops are – as well as services like banks and restaurants.

**b** Students should name their nearest high street.

**3 a** A chain store is one of a group of similar shops that are owned by the same company.

**b** An out-of-town shopping centre is a large complex which is located on the outskirts of a city / town, close to good road networks. These often have shops, cafes, restaurants and other amenities all on one site, with ample car parking.

**4 a** Online shopping is likely to be more convenient than instore shopping for most people because they can shop from their own home or from anywhere, using their smartphone. This is quicker and cheaper than going to a high street or an out-of-town shopping centre. However, some people who live in urban centres may pass shops regularly, so instore shopping could be quite convenient for them.

**b** Online shopping provides more choice than instore shopping, as consumers can use search engines to find the exact product they want, rather than looking in several stores. Many of these products come from abroad and may not even be available in physical stores in the UK.

**5** Answers will vary but all should consider how commercial units on the high street could be used in a future where physical shops are less common, eg gyms, communal workspaces. A labelled sketch would be a good way for students to present their ideas.

**6 a** Advantages of working from home: no time wasted commuting; no cost of commuting; no need to spend money on work attire; more flexible hours; reduced impact on the environment.

**b** Disadvantages of working from home: loneliness; lack of community; workers must have reliable Wi-Fi; back problems from using inappropriate furniture; distractions from family members and pets; the additional cost of utilities.

**7** This could be an interesting debate topic. Some students may prefer the routine of school, others may prefer some flexibility which allows them to learn from home. Discussion could focus on equality. Would students who have their own bedroom and a desk be able to make faster progress than students who share their space? Would it be possible for schools to provide laptops for students who don't have them?

**8** Most students will suggest that technology helped keep the economy afloat during the pandemic, particularly in the UK due to good broadband networks which have enabled people to work from home. Using the Internet, many people were still able to log into their work network to access files, and may have used Teams and Zoom to conduct business meetings. On the other hand, if technology was not so advanced, people may not have been asked to work from home, so the impacts on the economy in a scenario with less technology are hard to predict.

# So what kind of work will *you* do?



## About this unit

This unit looks at what kind of work might be available when students have to start earning a living.

## Key points

- Three factors could affect students' future working life: technology, globalisation, and the government.
- Artificial Intelligence (AI) will impact on the work that we do in future.
- The gig economy is growing. It uses online platforms for people to find work – in 2019 4.7 million people in the UK did some gig work.
- Green jobs are likely to grow, as the UK must be carbon-neutral by 2050. Green jobs are those that help to protect the environment and other species; limit climate change, and reduce the harmful impacts of climate change.

## Key vocabulary

Industrial Revolution, Information Technology (IT) Revolution, Artificial Intelligence (AI) Revolution, gig economy, online platform, carbon-neutral, carbon capture and storage (CCS)

## Skills practised in 'Your turn'

- Literacy skills: q2, q5, define
- Thinking skills: q1, q8, q9, use own knowledge; q1, q8, q9, explain; q2, q5, give examples; q3, q6, suggest advantages; q4, q6, suggest problems; q7, q9, create own definition; q10, design an ideal job

## Unit outcomes

By the end of this unit most students should be able to:

- explain the terms given in 'Key vocabulary' above
- explain how technology, globalisation, and the government might affect their future working life
- give an example of AI
- explain how the gig economy works, and give two benefits and two disadvantages of it for gig workers
- give four examples of green jobs and explain why they are green jobs

## Ideas for a starter

1. Write 'Technology', 'Globalisation', and 'The Government' on the board. Ask: How might these three things affect the work you will do in the future?
2. Ask: What does the term *green jobs* mean? Who can give me some examples of green jobs? Record students' ideas on the board and come back to them at the end of the lesson.

## Ideas for plenaries

1. 'The gig economy is better for workers as it gives them more freedom.' Discuss.
2. Use 'Your turn' question 8 as a plenary activity.

## Further class and homework opportunities

Suggestions 21 and 22 on page 83 of this book

*geog.3 workbook,* page 30

**kerboodle**

### Resources

3.8 You decide! – Students rank the criteria for choosing a job.

3.8 Quiz time! – Students complete a quick quiz about future jobs.

3.8 Foundation worksheet – Students predict what types of jobs will be available in the future, and examine how AI and the UK's goal to become carbon neutral will influence this.

3.8 Extension worksheet – Students evaluate how the government will influence jobs available in the future and whether the AI Revolution will bring only benefits.

### Assessment

3.8 End-of-lesson test – An assignable, auto-marked formative assessment.

## Answers to 'Your turn'

**1** The Internet has made remote working possible. People can connect to their colleagues in other countries via email, Zoom, etc. Language barriers are becoming less important due to advancements in translation software.

**2** AI is when machines are programmed to do jobs that usually need human intelligence.

**3** Sensors used by robots are much more efficient than the human senses. These sensors can be set to specific levels to detect minute differences, not perceptible by humans.

**4** AI is already being used to do jobs once done by humans, so unemployment could become an issue. AI may not always be used for ethical reasons. Some students may comment that robots gaining consciousness is a common sci-fi theme – if this happened, how would robots treat humans?!

**5** In the gig economy people use an online platform to find short / temporary jobs, rather than permanent ones. For example: Uber drivers, Just Eat delivery workers.

**6** **a** Benefits for workers: gig work is flexible, which suits some people. For example, if they have another job, they can work around it; or if they have children, they can fit in work around school hours and school holidays.

**b** Disadvantages for workers: no benefits such as paid holidays or sick pay; no pension; work is not guaranteed; wages may be low because some gig jobs are competitive.

**7** A green job is one which is focused on the environment. Some are linked to climate change; others are related to protecting biodiversity.

**8** **a** Installing solar panels is a green job because solar panels produce renewable energy. If we use more renewable energy, it means we can burn less fossil fuels, which are contributing to climate change.

**b** Inventing a way to clear plastic rubbish from the sea is a green job because the oceans are part of our natural world. Keeping them clean protects marine life such as fish, turtles, sea birds, and corals.

**c** Organising protests to ban pesticides that protect bees is a green job because bees provide ecosystem services and are integral to our survival on Earth. Bees pollinate crops, flowers and trees. The process of pollination fertilises plants and allows them to reproduce.

**9** **a** Carbon neutral means adding no carbon dioxide to the atmosphere overall; if any is being added, it is cancelled out by planting trees.

**b** Farmers can help by restoring soil health; reducing the use of pesticides to protect bees; growing vegetables instead of rearing animals; growing crops that can be used for biofuel; rewilding land which they don't use.

House builders can help by designing houses that: are energy efficient, with insulation to reduce heat loss; are fitted with solar panels to generate their own electricity; are linked to cycle networks so that residents are able to reduce their use of cars.

Car factories can help by: producing cars that are hybrid or electric; using renewable energy to power their machinery; incentivising their employees to cycle to work and / or use public transport.

**10** Encourage students to consider their passions and their talents, and to design a job which doesn't already exist.

# Earning a living



## Skills practised in check questions

- Geographical skills: q1, q2, q4, interpret a photo; q3, interpret a line graph
- Numeracy skills: q7, construct a pie chart
- Literacy skills: q1, q3, q4, classify; q2, q5, define; q2, unjumble a word
- Thinking skills: q1, make a link; q1, q2, q4, q7, explain; q2, q3, q5, give reasons; q4, q5, give examples; q6, draw a spider map; q6, consider factors relating to future jobs; q7, imagine a scenario; q7, make proposals

## Answers for check questions

**1 a** This person is working. This is entertainment and therefore part of the tertiary (service) sector. It can also be considered part of the informal sector.

**b** Students can make a link between any of the materials in the photo that have been extracted as part of the primary sector. The rock which has been used for the building or the paving stones is a likely choice; or the metal used for the railings / statue.

**c** Cinemas provide entertainment, so they are part of the tertiary sector.

**2 a** Traditionally, the high street is where the main shops are in an urban centre – as well as services like banks and restaurants.

**b** A chain store is a series of shops owned by the same company, selling similar goods. Coffee shops such as Costa are also chains.

**c** Many chain stores get their clothing made in countries like China, Bangladesh and Vietnam, because wages there are lower than in the countries where their headquarters are based. This means that costs of production are cheaper, and clothes can be sold at lower prices, which leads to higher profits overall.

**d** Globalisation

**e** High street shops have been struggling for decades due to the popularity of out-of-town shopping centres and the rise of online shopping.

**3 a** **i** About 6.7 million **ii** About 2.6 million

**b** The trend shows a decline in the number of people employed in manufacturing

**c** The UK lost industry during that period because: it became cheaper to import goods from abroad; minimum wages in the UK's manufacturing sector were too high to compete with wages in newly emerging economies such as China; coal mines closed down; the UK began to focus on the quaternary sector, which includes IT, scientific research, and finance

**d** The UK excels in high-value manufacturing, such as aerospace and pharmaceuticals

**e** 8%

**4 a** Cranes are being used to build a wind turbine.

**b** Secondary

**c** Wind turbines generate renewable energy, which is part of the green sector. If the UK builds more renewable energy sites like this, it will be less reliant on fossil fuels, which contribute to climate change.

**d** Two 'green jobs' should be suggested. For example: environmental biologist, climate scientist.

**5 a** The gig economy is where you use an online platform to find short temporary jobs, rather than permanent ones.

**b** Two gig economy jobs should be suggested. For example: delivery worker, taxi driver, plumbing, graphic design, legal work.

**c** Students should suggest one reason why people in the 16–34 age group are more likely to be involved in the gig economy. For example: some have not yet progressed into careers; some have young children and the flexibility suits them; some are students and can fit in work around their studies; some have struggled to find full-time jobs due to the competitive job market.

**6** Students should produce a spider map that shows factors they predict will affect jobs in the future. Suggestions: climate change; artificial intelligence; globalisation; changes in government policies; the rise of online services; new inventions.

**7 a** Students should create their own pie chart. They should use a compass and a protractor and they should calculate the angles before adding the segments which represent the percentages.

**b** Answers will vary. Students are likely to suggest that we need more jobs in the quaternary sector because these generate a high income and they utilise the technology that has become integral to our lives in richer countries; some may focus on an increase in secondary sector jobs to reverse the effects of de-industrialisation; some may focus on an increase in primary sector jobs such as farming which could improve the UK's food security and help to reduce the impacts of climate change related to food miles.

**c** Answers will vary. Students should focus on how to promote jobs in some sectors and consider how people may re-skill themselves if the job market changes.

To promote jobs in the primary sector, ministers could promote locally grown food; for the secondary sector, ministers could encourage apprenticeships in construction; for the tertiary sector, they could spend more on public services such as the NHS; for the quaternary sector, they could encourage young people to get higher education qualifications in IT, finance, and research.

# Further suggestions for class and homework

Many of these suggestions are addressed to your students. Where research or further resources are needed, the Internet will almost certainly provide the answer.

The suggestions are graded *, **, *** according to level of difficulty. Some are suitable for all levels, and can be differentiated by outcome.

## What kind of work do we do in the UK?

1 **The UK at work** Display images of all the different types of work people do in the UK, arranged by sector. Label each image with its job title (not the subject's name). You could get images from magazines, papers, the Internet. */**

2 **It's phenomenal!** Choose a geographical phenomenon or feature. For example the weather, earthquakes, rivers, a mountain. Then write a list of jobs connected with it! Can you find jobs in more than one sector? */**/***

3 **Healthy and fit** Make a list of jobs to do with keeping us healthy and fit. Think widely. What sectors? **/***

4 **The economy in the news** A student chooses a simple newspaper article about the British economy, and reads it to the class. Another student stands up and summarises it. ***

## The UK's changing pattern of employment

5 **Make a sector meter** Make a dial like an employment structure pie chart – but where you can easily change the sizes of the three sectors. In cardboard – or use software? **/***

6 **The Industrial Revolution ...** started in Britain. Tell us about it, as a presentation, an illustrated essay, a poem, a song, a set of sound effects, or a short play! */**/***

7 **Fair pay?** Some people get enormously well paid for their work. Some earn a pittance. If you were in charge, what rules would you make, to make sure pay was fair? **/***

## Example: change in and around Doncaster

8 **Why there?** *Why* did the Romans build a fort where Doncaster is? Do some research, and tell us. **/***

9 **Why does Doncaster have coal?** Doncaster sits on the South Yorkshire Coalfield. Coal is made from the remains of lush vegetation which grew in tropical swamps in the Carboniferous period. So how did it get to Doncaster? (Hint: Chapter 1!) Create a wall frieze linking these early lush forests to the miners in the photo on page 54 of *geog.3*. (You will need to find out more about how coal formed.) ***

10 **Train enquiry** Where are the UK's trains and buses made now? Carry out an enquiry. **/***

## Why do employment patterns change?

11 **The British cotton industry** Research the history of Britain's cotton industry. Then draw a time line for it. Annotate your time line with text panels and photos. Include information about people whose inventions helped the industry to expand. And also about Britain's relations with India over cotton. (Mahatma Ghandi's cotton protest?) ***

12 **The UK needs more manufacturing** Hold a class debate on the issue. If the motion is upheld, write to the prime minister giving the class's views, and reasons. Be persuasive! **/***

13 **Work for young people** You are a government minister. It's up to you to ensure that school leavers in the UK have work. How will you do this? Decide, then write an action plan. ***

## A closer look at globalisation and the UK

14 **Design a poster ...** to show why globalisation has speeded up. Make it colourful, and suitable for someone in year 7. */**/***

15 **G7 tax deal** In 2021 the G7 countries reached a deal to make transnational companies pay more tax. Find out: who the G7 are; what the deal was; which companies have been affected. **/***

## Is your mobile linked to globalisation?

16 **Where does the money go?** Miners and factory workers in China don't earn much. We pay a lot for iPhones. Where does the money go? ***

17 **Shareholders** 'Public' companies like Apple are owned by shareholders. What are these? How does it all work? ***

18 **Choose another item of yours** For example jeans, a bike. Find out which sectors – and if possible, which countries – were involved in bringing it to you. See if you can show the results in an exciting way. **/***

## Employment and the shock of COVID-19

19 **Choose a business ...** that used to be on the high street, but was forced to close by COVID-19. Imagine you are the owner. Explain what happened. Have you re-opened? How has your business changed? Tell us in any way you want – a story, podcast, poster. */**/***

20 **Crossword** Make up a crossword for some of the terms used in this unit. Give it to a partner. */**/***

## So what kind of work will *you* do?

21 **Play Taboo** Students work in small groups, decide on a green job and then on taboo words they can't use to describe it. One of the group describes the job, without using the taboo words. The class guesses. */**/***

22 **Carbon neutral now!** Can we wait until 2050 to be carbon neutral? Hold a class discussion; summarise your arguments; send an email or tweet your conclusions to the prime minister. */**/***

# 4 International development

## About this chapter

### Addressing the KS3 Programme of Study

- Understand, through the use of detailed place-based exemplars at a variety of scales ... human geography relating to international development
- Develop greater competence in using geographical knowledge, approaches and concepts ... in analysing and interpreting different data sources

### About the content

- The chapter starts by looking at inequality and poverty, what development means and how it is measured, and uses Malawi as a case study.
- It investigates the reasons for the development gap between countries, and the impacts of globalisation.
- It looks at how people migrate to escape poverty – a topic of continuing relevance.

## The big picture

These are the key ideas in this chapter.

- Development is a process of change that improves people's lives.
- There is a big gap in development between richer and poorer countries, and millions of people live in extreme poverty.
- Globalisation can help countries to develop, but there are downsides.
- We can put an end to extreme poverty.

## The chapter outline

Use this to give students a mental roadmap for the chapter.

4 **International development** The chapter-opening photo in the students' book is an important resource. See opposite for using it.

4.1 **An unequal world** How the world's wealth is shared; poverty, water supply, and sanitation

4.2 **So ... what is development?** Comparing a highly developed and poorly developed country, and obstacles to development

4.3 **How is development measured?** Development indicators and HDI

4.4 **How is Malawi doing?** The level of development in Malawi, and its links with the UK

4.5 **How did the development gap grow? – part 1** Historical, geographical, and social reasons why some countries are more developed than others

4.6 **How did the development gap grow? – part 2** Further reasons why poor countries find it hard to climb out of poverty

4.7 **Globalisation, development, and TNCs** How TNCs can speed up a country's development, plus the downsides

4.8 **COVID-19 and vaccine inequality** The pandemic was global, but richer countries were first in the queue for vaccines

4.9 **Escaping from poverty** How people risk everything to find a better life

4.10 **Can we put an end to poverty?** How poorer countries can help themselves, and how richer countries, NGOs, and technology can help

4 **International development check questions** A page of questions to check knowledge, understanding, and skills across the chapter, with an opportunity for extended writing

## Objectives and outcomes for this chapter

| Objectives | Unit | Outcomes |
|---|---|---|
| Most students will understand: | | Most students will be able to: |
| • that there is great inequality in the world | 4.1 | • define inequality and extreme poverty; give reasons why poverty may increase in many countries |
| • that development is a process of change to improve people's lives, and that countries are at different stages of development | 4.2 | • give examples of conditions you would expect in a highly developed, and poorly developed, country |
| • that different aspects of development can be measured, and that HDI includes three aspects of development | 4.3 | • give five examples of development indicators, and describe how HDI varies around the world |
| • that Malawi is one of the world's poorest countries | 4.4 | • describe Malawi's location; explain how different factors could have held back its development |
| • that there are reasons for the big development gap between rich and poor countries | 4.5, 4.6 | • explain how different reasons (historical, geographical, social, conflict and corruption, export type, and lack of industry) have held back development |
| • that globalisation is largely driven by TNCs, and that TNCs can help a country develop | 4.7 | • give two examples of TNCs; give arguments for and against TNCs in low income countries |
| • that the COVID-19 pandemic exposed global inequality | 4.8 | • suggest three reasons why poorer countries found it hard to deal with the pandemic; describe how richer countries were first in the queue for vaccines |
| • that many people are driven to escape from poverty and make a better life, by moving to another country | 4.9 | • explain that many migrants head for other countries without permits or visas; give reasons why people are willing to risk danger to migrate |
| • that poorer countries can help themselves to develop, and that richer countries, NGOs and technology can help | 4.10 | • explain how different actions can help lift people out of poverty; give an example of top-down development and bottom-up development |

These tie in with the outcomes for each unit in this teacher's handbook.

## Assessment for this chapter

There are assessment materials for this chapter on *geog.3 Kerboodle Lessons, Resources, & Assessment*. They include interactive auto-marked end-of-lesson and chapter-level tests, a printable exam-style end-of-chapter paper (with a mark scheme), and a printable self-assessment checklist. See also the notes on pages 14–17 of this book.

## Getting ready for this chapter

There is a scheme of work for this chapter, plus a glossary and glossary worksheet, on *geog.3 Kerboodle Lessons, Resources, & Assessment.*

## About the chapter-opening photo

Ask students:

- What is this girl doing?
- What is the connection between this photo and development?
- What image would you have chosen to illustrate development?

The photo shows a tightrope walker on the beach in Goa, India, entertaining tourists. Like a country's development process, tightrope walking involves goals, focus, balance, overcoming obstacles, and trying not to get pushed off track.

# 4.1 An unequal world

Help at a glance

## About this unit

This unit explores just how unequal the world is.

## Key points

- There is great inequality in the world – many live in extreme poverty, while some people are extremely rich.
- Poverty is where people do not have sufficient resources to meet their basic needs; extreme poverty is where people have resources worth less than $1.90 a day to live on.
- Forty-four per cent of the world's population live in poverty, or extreme poverty; 33% do not have clean, safe water at home; only 45% have safe toilets.
- Although poverty had been falling globally, COVID-19 caused it to rise again.

## Key vocabulary

inequality, poverty, extreme poverty, poverty lines

## Skills practised in 'Your turn'

- Geographical skills: q1, interpret photos; q4, interpret diagrams
- Literacy skills: q3, define
- Numeracy skills: q4, calculate
- Thinking skills: q1, comparison to students' own life; q2, give reasons; q5, give an opinion; q5, explain; q6, apply own knowledge; q6, suggest reasons

## Unit outcomes

By the end of this unit most students should be able to:

- explain the terms given in 'Key vocabulary' above
- describe four aspects of the life of a person living in poverty in a poor country
- give the percentage of the world's population living in poverty and extreme poverty
- suggest two reasons why climate change is likely to increase poverty in many countries

## Ideas for a starter

1. With books closed, show an image of Earth from space, on the board. Write the heading 'An unequal world'. Ask students what they think this means. They could create a graffiti wall of ideas they come up with.
2. Ask students to close their eyes and imagine that they're not born yet. Will they have any control over where in the world they will be born? Will where they are born affect their life? Why? How?
3. Hands up who thinks the UK is wealthy. Can you name any countries you think are wealthier than the UK? What about countries that are far less wealthy?

## Ideas for plenaries

1. Use 'Your turn' question 2 as a plenary activity.
2. Ask: Do you think we have a responsibility to do something about world poverty?
3. Do you find this topic interesting so far? Why?

## Further class and homework opportunities

Suggestions 1 and 2 on page 107 of this book

*geog.3 workbook*, page 31

**kerboodle**

### Resources

4.1 You decide! – Students order statements about development.

4.1 Word scramble – Students complete a crossword of key terms about poverty.

4.1 Foundation worksheet – Students use infographics on poverty, water supply, and sanitation to complete facts and write a diary entry imagining life under the poverty line.

4.1 Extension worksheet – Students analyse inequality in the world and explain why the wealth of a country does not reflect equally upon its citizens.

### Assessment

4.1 End-of-lesson test – An assignable, auto-marked formative assessment.

## Answers to 'Your turn'

1 Students should write four paragraphs about their own life as a comparison to the lives of people in a poor rural village. These should focus on: kitchen / household chores; water supply; toilet facilities; money and / or medical costs.

2 Students are likely to suggest that it would be difficult for people in this situation to improve their life because their time is consumed with basic survival tasks, such as fetching water and firewood. A discussion about what it means for a person to 'improve their life' may be useful. Examples are likely to include: earning more money, home improvements, getting an education, and accessing better medical care. However, personal growth can be spiritual, or it may relate to being part of a community, so people living in poor rural areas may have some scope to improve these aspects.

3 **a** When wealth and access to services are not shared equally.

**b** When people are living on less than $1.90 a day.

4 44%

5 Students should select statistics or statements and explain why they are shocking. For example, some may be surprised that one person (1% of the world population) has half of all the wealth. Students may wonder why this wealth is not shared out so that those in poverty do not have to suffer.

6 Students should give two reasons why climate change may increase poverty. For example: rising sea levels may displace people from low lying land; droughts and floods could become more common which would impact farmers; pests and diseases may affect more crops and lead to famines; more tropical storms could cause coastal flooding and high winds could damage property; conflict could result between countries due to rising numbers of climate refugees and water insecurity.

# 4.2 So ... what is development?

Help at a glance

## About this unit

This unit explores what development means, how poverty and development are linked, and how countries are at different stages of development.

## Key points

- Development is a process of change that improves people's lives.
- People in highly developed countries are likely to have a higher quality of life with more opportunities and choices than those in poorly developed countries.
- There is a big development gap between the world's richest and poorest countries.
- There are a number of obstacles that can slow down, halt, or reverse development.

## Key vocabulary

level of development, quality of life, development gap

## Skills practised in 'Your turn'

- Geographical skills: q6, interpret a photo
- Literacy skills: q3, q4, define
- Thinking skills: q1, q2, consider significance of factors; q1, rank and justify; q1, decide; q1, give evidence; q2, q5, give reasons; q3, write a summary; q3, q5, interpret a drawing; q5, deduce; q6, explain

## Unit outcomes

By the end of this unit most students should be able to:

- explain the terms given in 'Key vocabulary' above
- give at least three conditions they would expect to find in a highly developed country
- give at least three conditions they would expect to find in a poorly developed country
- say that countries are at different stages of development and that there is a development gap between the richest and poorest countries
- explain how war can reverse development

## Ideas for a starter

1. With books closed, write on the board: *Development – change that improves people's lives*. Ask students for an example of something that would:
   - change their lives for the better
   - change their local area for the better
   - change the country for the better
   - change the whole world for the better

   Look at the responses. Are they about money? Is development just about money? Would a person in a poorly developed country give the same kind of answers?
2. Recap Unit 4.1. Ask: What percentage of the world's population lives in poverty and extreme poverty? What percentage doesn't have a safe water supply, or access to a safe toilet? A high percentage of the world's poorest people live in poorer countries and their lives are shaped by the level of development in their country.

## Ideas for plenaries

1. Discuss development in the UK in the light of 'Your turn' question 5.
2. Tell the class one big insight that you had today.

## Further class and homework opportunities

Suggestions 3 and 4 on page 107 of this book

*geog.3 workbook*, page 32

**kerboodle**

### Resources

4.2 Sort it out! – Students categorise features of development.

4.2 You decide! – Students order indicators of development.

4.2 Foundation worksheet – Students suggest ways that their local area could be developed and describe the obstacles that advanced countries still face.

4.2 Extension worksheet – Students rank their top five aspects of development, justifying their choices.

### Assessment

4.2 End-of-lesson test – An assignable, auto-marked formative assessment.

## Answers to 'Your turn'

**1** **a** Answers will vary

**b** **i ii** Answers will vary

**c** The UK is highly developed. Evidence: safe water, sanitation, education and health care is available to everyone; high quality infrastructure; high % workers in service sectors; most people live in urban areas; low fertility rate; high median age of population.

**2** Answers will vary

**3** **a** The steep and rocky slope, with a drop on each side, suggests that the road to development is not an easy one. You have to take care. You could drop off.

**b** Quality of life is the level of comfort and well-being a person enjoys.

**c** Students could include these points: the road to development is a tough one; every country is at a different stage of development; some are much more developed than others; there are many reasons why countries lag behind in development; their level of development can change over time.

**4** An emerging country is one which is developing fast due to less reliance on farming and the growth of industry.

**5** **a** As a developed country, the UK can also be considered 'high income' and 'advanced.' People in the UK have a high quality of life.

**b** The UK's position could change if its GDP decreased while the GDP of other countries continued to rise. This could result from other countries expanding their quaternary sector, or if there was a war and the UK lost some of its wealth. A pandemic, eg COVID-19 could impact on development, and children could lose years of school. Climate change could also have an impact.

**6** The spider map could look something like the one below.

# 4.3 How is development measured?

Help at a glance

## About this unit

In this unit students will learn how we can compare countries at different stages of development.

## Key points

- Development indicators help to show how developed countries are.
- Development indicators can be social (eg number of doctors per 100 000 people), economic (eg GNI per person [PPP]), or demographic (eg life expectancy).
- The human development index (HDI) looks at three aspects of development: life expectancy, access to education and standard of living (measured by GNI per person (PPP).
- The HDI value rises as a country develops. But it can also fall.

## Key vocabulary

data, development indicator, life expectancy, GNI per person (PPP), human development index (HDI)

## Skills practised in 'Your turn'

- Geographical skills: q1, interpret a drawing; q3, q6, interpret a map; q3, compare maps
- Literacy skills: q1, use a glossary
- Numeracy skills: q2, q3, q4, interpret HDI scale
- Thinking skills: q1, interpret a drawing; q1, match statements to indicators; q1, make comparisons; q2, comprehension; q2, q4, deduce; q5, explain; q6, suggest reasons

## Unit outcomes

By the end of this unit most students should be able to:

- explain the terms given in 'Key vocabulary' above
- give examples of what different development indicators measure
- describe how HDI varies around the world
- deduce three things about a country with a high HDI; and deduce three things about a country with a low HDI
- give three reasons why HDI can fall

## Ideas for a starter

1. With books closed write this on the board: *On average, people in the UK live to 81, and people in Mali, in Africa live to 62*. Ask for suggestions to explain the difference. Could the ages be linked to development? Elicit the idea of life expectancy as a development indicator.
2. Ask students to imagine themselves as adults, unable to read or write because they spent little time at school as a child. Elicit the idea that we can measure literacy rates, and use adult literacy rates as a development indicator.

## Ideas for plenaries

1. Show world maps for other indicators – try searching the Worldmapper website where territories are resized according to the subject. You could use the map for underweight children, illiterate adults, infant deaths, etc.
2. Ask: Does the value of a development indicator for a country stay the same, year after year?

## Further class and homework opportunities

Suggestions 5 and 6 on page 107 of this book

*geog.3 workbook*, page 33

**kerboodle**

### Resources

4.3 Sort it out! – Students identify indicators of development.

4.3 Word scramble – Students complete a crossword of key terms for development indicators.

4.3 Foundation worksheet – Students identify whether a country is developed or developing, and colour and complete a graph using adult literacy rate figures.

4.3 Extension worksheet – Students rank development indicators, consider the value of the human development index, and examine the global pattern of HDI values post COVID-19.

### Assessment

4.3 End-of-lesson test – An assignable, auto-marked formative assessment.

## Answers to 'Your turn'

1

| Question | Indicator which answers it | |
|---|---|---|
| 1 Do children get enough to eat in that country? | % of children under age 5 who are underweight | H |
| 2 Does everyone have a clean, safe water supply? | % of the population with access to clean safe water | L |
| 3 Can people expect to live long lives? | life expectancy (years) | L |
| 4 What are the chances of dying when you are only little? | under-5 mortality rate (%) | H |
| 5 Is it easy to see a doctor if you fall ill? | number of doctors per 100 000 people | L |
| 6 Can everyone over 15 read and write? | adult literacy rate (%) | L |
| 7 Are people wealthy there, or poor? | GNI per person (PPP), ($) | L |

2 **a** Human development index

**b** Any three from: high life expectancy; adults had good access to education / many years schooling; children can expect the same; there is a good standard of living / high GNI per person. Students may refer to other indicators, but it should be noted that these are not part of the HDI. For example: access to clean water and sanitation; doctors per person; child mortality rate; % of underweight children.

3 **a** Africa

**b** **i** Most likely answers are: Niger, South Sudan, Central African Republic. Any represented by the lightest green colour are also applicable.

**ii** Any three from: Angola, Zambia, Zimbabwe, Madagascar, Tanzania, Uganda, Sudan, Nigeria, Mauritania, Pakistan. Any represented by the second-lightest green colour are also applicable.

**iii** Any three from: Australia, Austria, Belgium, Canada, Denmark, Germany, Finland, France, Iceland, Israel, Japan, Luxembourg, the Netherlands, New Zealand, Norway, South Korea, Sweden, Switzerland, UK, USA.

4 Any three from: low life expectancy; adults had poor access to education / few years schooling; children can expect the same; there is a low standard of living / low GNI per person.

5 Climate change could create challenges for a country which are expensive to tackle, and would affect a country's GNI. For example, rising sea levels could require expensive flood defence schemes, or relocation of entire urban areas. Impacts on agriculture could reduce GNI if crops fail. The social elements of HDI could also be affected by these challenges if lack of food causes a famine which may lead to deaths; or if reduced income meant that people could no longer afford to send their children to school.

6 **a** X is Greenland which has only a very small population due to its extreme climate. Residents live in remote places and data may not be collected there.

**b** Y is Somalia which has an unstable government. Data may not have been collected due to conflict or lack of funds / organisation.

# 4.4 How is Malawi doing?

Help at a glance

## About this unit

Malawi is one of the world's poorest countries. This unit looks at its geography, history, economy, and some development indicators, plus a day in the life of a primary-school student. The aim is to help students understand what poverty means.

## Key points

- Malawi is a small landlocked country in East Africa.
- It has a population of over 19 million.
- It is one of the world's poorest countries – around 51% of the population live in poverty.
- It was once under British rule, but gained independence in 1964.

## Key vocabulary

Malawi, landlocked

## Skills practised in 'Your turn'

- Geographical skills: q1, describe location; q4, interpret and compare pie charts
- Literacy skills: q5, use a glossary
- Numeracy skills: q1, q3, subtraction; q2, division
- Thinking skills: q3, suggest reasons; q4, q6, compare; q5, explain

## Unit outcomes

By the end of this unit most students should be able to:

- explain the term *landlocked*
- describe where Malawi is
- compare development indicators for the UK and Malawi
- recognise that life for a poor child in Malawi is very different from children's lives in the UK

## Ideas for a starter

1. With books closed, display a set of images of Malawi on the board. Include photos of rural villages, primary school classrooms, cities, landscapes, farming, fishing, people in tertiary jobs, and wildlife. Then ask for students' impressions of Malawi. Could different photos give completely different impressions?
2. With books closed, give students five or six minutes to find Malawi in an atlas, and write at least six geographical facts about it. Remind them to look at different world maps, including climate. Students can work in pairs. Collect answers on the board.

## Ideas for plenaries

1. Ask: What are the main things you notice about Malawi, from the map on page 72?
2. Ask questions about the table of indicators. For example: What impact might the number of average years of education have on people in Malawi?
3. If you did not use starter 1: display images of Malawi from the Internet.
4. Ask: Would you like to go on a school trip to Malawi? Why? / Why not?
5. Ask: Are you wealthy, compared to Sephora? If you could help Sephora by doing one thing, what do you think she would want you to do?

## Further class and homework opportunities

Suggestions 7 and 8 on page 107 of this book

*geog.3 workbook*, page 34

**kerboodle**

### Resources

4.4 Where in the world? – Students complete a table of development indicators for Malawi and the UK.

4.4 You decide! – Students order factors which have hindered Malawi's development.

4.4 Foundation worksheet – Students retell the history of Malawi and complete four drawings showing what life is like there.

4.4 Extension worksheet – Students draw and write the story of Sephora's day, described on page 73 of *geog.3*, in Malawi.

### Assessment

4.4 End-of-lesson test – An assignable, auto-marked formative assessment.

## Answers to 'Your turn'

1 **a** It is a landlocked country in east Africa. It is a long thin country, bordered by Tanzania, Zambia, Zimbabwe, and Mozambique. (Easy to remember – they all have *z* in their names.)

**b** 73 years

2 Over 44 times richer

3 A person in the UK will live 17 years longer than a person in Malawi, on average. This is because many people in Malawi have no access to clean, safe water and do not have enough nutritious food and there are not enough doctors. (On average there are four doctors per 100 000 people, compared with 281 in the UK.) But remember, life expectancy is an average value. In Malawi you will find many people older than 64. What reduces the average is that many infants die during childbirth, or soon after. AIDS is still a big problem. Around 1.1 million people have AIDS and many babies are born HIV positive. Death rates have fallen due to the progress in antiretrovirals, but thousands are still dying every year.

4 In the pie chart for Malawi, the largest sector is still primary. This is mainly farming and stands at 77%. In contrast, only 1.3% of people in the UK work in the primary sector. The UK's service sector (84%) is more than four times larger than Malawi's (19%). In the UK, the tertiary sector also includes the quaternary sector which incorporates research, finance and technology. Overall, in the UK, there are more jobs in sectors which are associated with higher wages.

**a** If Malawi had a coast, exporting its commodities and importing goods would be easier and cheaper. Commodities are transported to ports in Mozambique and Tanzania, which adds greatly to the cost of exporting. A coast is also good for fishing – but Lake Malawi is an excellent source of fish.

**b** 4.7 years in education is less than half of the UK's figure – 13.2 years. Students should consider what it would be possible to learn in 4.7 years and which things may be missed out. Without a good education, these children would be unable to apply for universities and not able to work in higher paid sectors, which will be reflected in Malawi's GNI.

**c** Rapid population growth over a short time span means that Malawi's population is very young. In the long run, this may be useful as these people will be able to work and contribute to the economy; but while they are young, they are dependent on their families. School resources may be stretched and class sizes are likely to increase, which will reduce the impacts of education even further.

5 Answers will vary – but this exercise should help to bring out the effect of poverty on people's lives.

# 4.5 How did the development gap grow? – part 1

Help at a glance

## About this unit

This unit looks at some reasons why some countries are more developed than others: historical, geographical, and social reasons.

## Key points

- Historical reasons why some countries are more developed: the Industrial Revolution was a key factor in helping the UK and other European countries to develop. Several European countries colonised other countries and exploited their natural resources to their own advantage.
- Geographical reasons: a good climate and fertile soil for farming, natural resources which can be traded, and ease of access for trading: all help a country to develop.
- Social: a well educated, skilled, and healthy workforce will help a country to develop.

## Key vocabulary

Industrial Revolution, colonies

## Skills practised in 'Your turn'

- Thinking skills: q1, q2, q3, explain; q3 give reasons

## Unit outcomes

By the end of this unit most students should be able to:

- explain the terms given in 'Key vocabulary' above
- give at least one example of a historical factor, a geographical factor, and a social factor, which may have helped a country (such as the UK) to develop
- by contrast, give examples of factors which can hold back development

## Ideas for a starter

**1** With books closed, display a variety of images of the UK which indicate a developed country, but also to show that there is inequality.

Say that the UK is one of the world's more developed countries. How did it get to be like this? Students draw a spider map on the board, or on paper, with their suggestions.

**2** Draw shapes representing two countries of roughly the same size on the board, with placards as on the right. Ask: Which country is likely to develop faster? Why? (You could repeat with different labels.)

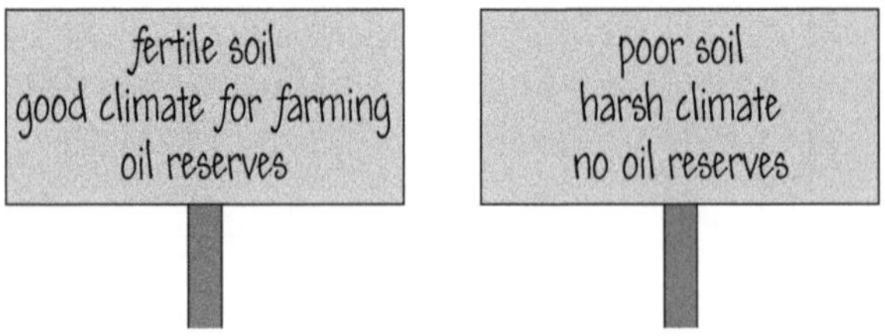


## Ideas for plenaries

**1** Any of the 'Your turn' questions could be used as plenaries.

**2** If you used starter 1 students should go back to the spider map, and modify or add to it.

**3** True or false? Call out a set of statements. Students answer with a show of hands, or hold up pre-prepared True / False cards. For example: The UK was always one of the world's richest countries. The UK will always be one of the world's most developed countries. Malawi will always be a poorly developed country. A country's level of development changes over time. Trade with other countries can help a country to develop.

**4** With books closed: Ask the students to tell you the three main things they have learned today. Ask them to tell you three other interesting, but less important, things they have learned.

## Further class and homework opportunities

Suggestions 9–11 on page 107 of this book

*geog.3 workbook*, page 35

**kerboodle**

### Resources

4.5 Sort it out! – Students identify factors which contribute to development.

4.5 You decide! – Students order the factors for development.

4.5 Foundation worksheet – Students label two spider diagrams with possible factors that could make a country rich or poor.

4.5 Extension worksheet – Students use a map to explain what the development gap will look like in 50 years.

### Assessment

4.5 End-of-lesson test – An assignable, auto-marked formative assessment.

## Answers to 'Your turn'

**1** **a** A poor education system holds back development because without a good education, people are not able to attend university and find jobs in the highest paid sectors such as engineering and medicine. Many bright youngsters will have no option but to stay and work as farm labourers in the small rural settlements where they grew up. This might suit some, but it does deter development, because countries which export primary resources as their main source of income have low GNIs.

**b** During colonisation, richer countries took resources and people from the countries they controlled, but did not spend much money improving education, health provision or infrastructure. After independence, these ex-colonies struggled to develop and there is now a wide gap between high income and low income countries.

**c** Malaria is a disease linked to bites from infected mosquitoes. It causes a high fever and flu-like symptoms and many die from it. High prevalence of disease slows development as people cannot work if they are unwell or if they are taking care of sick children.

**d** Being hemmed in by mountains, and hard to reach, means trade will be difficult. The country will find it hard to export crops and other goods it produces, and import what it needs. So it won't earn much from trade. This will hinder development.

**2** Students might decide that all could be put right / improved.

**a** The government could provide more money for education and employ more trained teachers.

**b** In theory, the legacy of colonisation can be left behind as countries can develop their own economy and services over time. However, this is difficult as many poorly developed countries have large debts.

**c** Malaria can be treated, and mosquito nets / insect repellant can be used.

**d** You can blast out routes into a country to improve access, and you can build airports.

**3** Reasons could include: it's an island, and sea access helped trade with other countries, and fishing. Its fertile land and temperate climate (because of its latitude) helped agriculture. The Industrial Revolution made it the workshop of the world. Access to the Atlantic Ocean led to the colonisation of other countries, and the British Empire. This provided raw materials and trade outlets, and made some people very wealthy. A good education system developed over the centuries, providing people with high levels of skill. Advances in medicine and the development of the health service led to a reasonably healthy population, able to work.

# 4.6 How did the development gap grow? – part 2

Help at a glance

## About this unit

This unit looks at other factors which help to keep poor countries poor.

## Key points

- Conflict, and corruption, are two other social factors which affect development. They can help to keep poor countries poor.
- Many poor countries depend on exporting just one or two commodities (such as sugar, or coffee). This is risky since the world price, or world demand, can fall at any time.
- Lack of industry does not help either. If poor countries had the factories, they could process their commodities themselves, and earn more. For example, turn cocoa into chocolate for sale.

## Key vocabulary

conflict, corruption, cash crops, commodities

## Skills practised in 'Your turn'

- Literacy skills: q1, q4, define; q3, draw a flow chart
- Thinking skills: q2, q3, q4, explain

## Unit outcomes

By the end of this unit most students should be able to:

- explain the terms given in 'Key vocabulary' above
- give examples of factors which hold back development
- explain how setting up factories could help poor countries to develop

## Ideas for a starter

1. With books closed, show images of conflict, or the result of conflict, from all over the world. Ask the class: How does conflict affect development?
2. With books closed, write CORRUPTION in big letters on the board, but jumbled. Say that this also affects development. Can anyone unjumble it? When unjumbled, ask what effect corruption might have on development.

## Ideas for plenaries

1. Ask: Do you think there is corruption in the UK?
2. Ask students how many commodities they can name (include cash crops).
3. Ask students to look at the pair of photos at the bottom of page 77 of their books. Why were those two photos put together? What points are this pair of photos trying to make?
4. Display sets of images showing how industry adds value to cash crops, by processing them. For example:

   oranges being harvested / bags of oranges in supermarkets / orange juice, marmalade

   cotton being harvested / cotton T-shirts, sheets, towels

   You could give prices if you know them.
5. Ask: Do you think it's important to learn about development around the world? Why do you think that?

## Further class and homework opportunities

Suggestions 12 and 13 on page 107 of this book

*geog.3 workbook*, page 36

kerboodle

### Resources

4.6 Word scramble – Students complete a crossword of key terms relating to development.

4.6 You decide! – Students rank the reasons why it's difficult for poor countries to escape from poverty.

4.6 Foundation worksheet – Students match up pairs of cards explaining why some poor countries are kept poor.

4.6 Extension worksheet – Students describe how certain factors contribute to keeping poor countries poor.

### Assessment

4.6 End-of-lesson test – An assignable, auto-marked formative assessment.

## Answers to 'Your turn'

**1 a** Corruption: dishonest conduct by people in official positions, usually to benefit themselves; for example they may accept bribes.

**b** Commodities: agricultural produce, and natural materials, taken from Earth and sold in bulk; for example, coffee, iron ore, oil.

**c** Adding value: processing something in a way that increases its value, so you can charge more.

**2 a** Corruption means that money meant for development often ends up in the private bank accounts of political leaders – who may spend it on lavish lifestyles. If money intended for new roads was stolen, this would reduce income from trade because transporting goods over poor roads takes longer and is less efficient. Roads are also used by people to get to work, schools and hospitals, and if these journeys take longer, people will have less time to participate in activities that lead to development.

**b** Depending heavily on a sole export such as cotton creates vulnerability to many shocks. For example crop disease, or a drop in the world price because other countries have produced a bumper crop, or because demand has fallen due to health fears (eg tobacco).

**c** If there are very few factories and almost everything is imported, this means that other countries are making all the profits from consumers. By contrast, if countries sell their own produce, the profits go back into the economy, particularly if people buy from small businesses, not large corporations that avoid tax. Income generated by businesses contributes to a country's GNI, which relates to development.

**3** You can export the cocoa beans, and leave it to other countries to add value and make good profits by processing them into chocolate. Or you could make the chocolate yourself and export it. That way you'll earn more from the export, and provide local people with work. So there'll be more money flowing in the economy, which will help development. (Then people will be more likely to afford nutritious food, and medicines; children will be more likely to continue at school; wells, solar power, and other improvements, are more likely to be made available.)

**4 a** Facilities such as water and electricity supplies, and road and rail networks, which help society to function

**b i** Students could touch on some of these: good roads are essential for businesses to work well; poor roads mean it is harder / slower for farmers and other producers to get goods to places where they can sell them, or to buy things they need (such as fertilisers); exporting and importing are harder, because of the difficulty in getting things to and from ports; it is harder for children to get to school, and for people to get to workplaces, markets, hospitals, and so on; everything takes more time and effort, and vehicles use more fuel and are more liable to damage; because of poor roads, there's less time, energy, and money for other things that would help development.

**ii** An unstable electricity supply would affect daily life. People use electricity for lighting, appliances and technology. Hospitals need it to take care of patients. Businesses need it to keep computers and Internet functioning. It is also vital in manufacturing and heavy industry.

In some cases, generators can be used to keep the supply running, but the cost of these is much higher than for electricity from the grid. Without reliable energy tasks take longer, and this slows development.

# 4.7 Globalisation, development and TNCs

## About this unit

In this unit students look at how TNCs drive globalisation, and bring both benefits and disadvantages to developing countries.

## Key points

- Globalisation is the process of making the world more connected through flows of trade, money, people and knowledge. It is driven by TNCs.
- Development in China took off in the 1980s when it invited foreign TNCs to set up. It is now the world's top manufacturing nation.
- Many TNCs from wealthy countries operate in lower-income countries in order to reduce costs, and be close to large markets.
- TNCs can help countries to develop, but there are disadvantages.

## Key vocabulary

globalisation, transnational corporations (TNCs)

## Skills practised in 'Your turn'

- Literacy skills: q1, define
- Thinking skills: q1, give reasons, give examples; q2, q3, choose and then justify; q4, imagine a scenario; q4, explain a viewpoint; q5, write a reply to an opinion

## Unit outcomes

By the end of this unit most students should be able to:

- explain the terms given in 'Key vocabulary' above
- give two examples of TNCs
- explain why TNCs from wealthy countries want to operate in lower-income countries
- explain why a government minister, and a mum with a young family in a low-income country would be in favour of TNCs
- explain why a doctor, and a young entrepreneur in a low-income country might be against TNCs

## Ideas for a starter

1. Ask: Who can remind me what globalisation is? Check back with Unit 3.4.
2. Ask different students to read out some of the points in favour of, and against, TNcs in developing countries from page 79 of the students' book, to help set the scene for this lesson.

## Ideas for plenaries

1. Use 'Your turn' question 4 as the basis for a class discussion. Students could make up their minds and write up their responses for homework.
2. Take a vote. Would students be prepared to pay more for their clothing to help the workers in developing countries who made it get fair pay and conditions?
3. Ask students to summarise what they have learned today in 35 words or less.

## Further class and homework opportunities

Suggestions 14 and 15 on page 107 of this book

*geog.3 workbook*, page 37

kerboodle

### Resources

4.7 Quiz time! – Students complete a quick quiz about TNCs and development.

4.7 Sort it out! – Students sort the advantages and disadvantages of TNCs.

4.7 Foundation worksheet – Students fill out a crossword puzzle on the topics of globalisation, development, and TNCs.

4.7 Extension worksheet – Students complete a writing frame responding to the statement that globalisation and TNCs can hinder development in developing countries.

### Assessment

4.7 End-of-lesson test – An assignable, auto-marked formative assessment.

## Answers to 'Your turn'

**1 a** A transnational corporation (TNC) is a company that has branches and operations in more than one country.

**b** The main reason for setting up in another country is to make as much profit as possible. This could be linked to cheaper raw materials, lower wages, or better access to markets.

**c** Students can list any two TNCs. For example Disney, Walmart, Shell, Amazon, Toyota.

**2 a** For a government minister, any of these options are relevant.

**b** For a mum with a young family, the most important benefits are likely to relate to her children, as parents often prioritise them. For example: the government can use the taxes to build schools, good wages would mean she could afford to send her children to school.

**3** Again, answers will vary for the government minister. For the mum with the young family, suggestions might be linked to low wages, as she would need to be well paid to afford education for her children; or issues affecting the environment as she would not want her children to grow up on a planet that has been wrecked by pollution and climate change.

**4 a** A doctor might be against TNCs because they are polluting the environment which is causing health issues. For example, toxic waste in rivers could get into the water supply and cause diseases; emissions from factories pollutes the air and affects people's lungs. The doctor may have experienced an increase of patients suffering from these health complaints. Conversely, a doctor might be in favour if TNCs mean more hospitals are built, and sanitation is improved.

**b** Someone trying to start their own factory would probably be against TNCs. Large foreign companies come and set up factories that compete with small local businesses so it can be difficult for them to make a profit. On the other hand, TNCs have caused mass rural-to-urban migration in some developing countries and there are many migrants arriving in cities in search of work, which may benefit the person who wants to start their own factory.

**5** This question will be a challenge for most students. Yes, TNCs move to developing countries to make a profit (either because wages are low, or because there is a natural resource to exploit). Local people will receive wages from the foreign companies, which may transform their lives. However, if a company pays less than a living wage, or destroys the environment, the host country may be better off without it.

# COVID-19 and vaccine inequality

Help at a glance

## About this unit

In this unit students will discover how the COVID vaccines were developed so quickly, and why richer countries were first in the queue for vaccines.

## Key points

- COVID-19 was spread by infected travellers, and initially spread fastest to richer countries.
- COVID-19 was declared a pandemic in March 2020 and few countries were prepared.
- Vaccine development was seen as a way to end the pandemic, and the first vaccines were ready in under a year (far faster than normal vaccine development).
- By the end of 2020, rich countries with just 14% of the world's population had booked over 50% of the expected vaccine supplies for 2021.
- COVAX – a scheme to share vaccines more fairly – was set up.

## Key vocabulary

pandemic, vaccine diplomacy, COVAX, mutate

## Skills practised in 'Your turn'

- Thinking skills: q1, q2, q5, explain; q3, q4, suggest reasons; q4, use your own knowledge; q6, give an opinion

## Unit outcomes

By the end of this unit most students should be able to:

- explain the terms given in 'Key vocabulary' above
- explain why air travel helped COVID-19 become a pandemic
- explain how vaccine development happened so quickly, and why most were developed in richer countries
- suggest reasons why vaccinating everyone in poorer countries is a challenge
- decide whether the COVAX scheme is a good idea

## Ideas for a starter

1. Show the image of the plane from page 80 of the students' book on the board. Ask: What is the link between this and COVID-19?
2. Ask different students to read out the speech bubbles from the diagram on page 80 of the students' book to set the scene for this lesson.

## Ideas for plenaries

1. Could a global scheme like COVAX work for other global issues, such as tackling the impacts of climate change? Have a class discussion.
2. Use any of the questions from 'Your turn' as plenary activities.

## Further class and homework opportunities

Suggestions 16 and 17 on page 107 of this book

*geog.3 workbook*, page 38

kerboodle

### Resources

4.8 Where in the world? – Students complete a paragraph about the spread of COVID-19.

4.8 Sort it out! – Students complete a description of the COVAX scheme.

4.8 Foundation worksheet – Students explain what COVAX is using increasingly fewer words and write a paragraph to persuade richer countries to donate money to COVAX.

4.8 Extension worksheet – Students write a report explaining why the COVID-19 vaccine was successful and outlining the problems faced by low-income countries in accessing vaccines.

### Assessment

4.8 End-of-lesson test – An assignable, auto-marked formative assessment.

## Answers to 'Your turn'

**1** Due to people travelling between infected countries by plane, the virus began to spread over long distances. This was mainly between the richer, more developed countries, because these people could afford to use air travel more frequently.

**2** **a** Closing borders posed a threat to tourism and trade and many governments opposed these measures, which some considered extreme. For example, countries which rely on tourism knew they would lose out financially due to travel bans. In most cases, trade has been allowed to continue, but restrictions at the borders has caused shortages. Businesses argued that international travel was important as part of their operations. Closed borders have also significantly affected migration patterns and millions of refugees have been stranded.

**b** Many countries were reluctant to shut down businesses due to the impact it would have on the economy. If people are not working and earning money it reduces GNI. Some countries paid furlough to people who were not able to work, which was expensive. Those countries which didn't lost out too, because unemployed people stopped spending.

**c** Closing schools was contentious because it meant students would lose out on their education. It also posed childcare issues for parents who work. Some countries knew it would be harder for students to work at home if households had no Internet access, and children did not have computers, desks or stationery of their own.

**3** Poorer countries have less money to spend on protecting people from the impacts of a crisis, so they struggled during the pandemic. For example, they could not afford to pay furlough and many who lost their jobs struggled to survive; in poorer countries many people work in informal jobs – such as street food vendors – and they lost income during the lockdowns; hospitals did not have enough beds, so ill people were not properly looked after; richer countries bought most of the vaccines so poorer countries could not protect their population.

**4** **a** Most vaccines were developed in richer countries because vaccine development is part of the quaternary sector, which is not part of the economy of all countries. Scientists working in research labs are highly skilled and need higher education qualifications. There are also high costs associated with vaccine development.

**b** In poor countries, many people are not officially registered so it may be hard to keep track of vaccinations. Plus, vaccines are expensive.

**5** **a** Poverty levels may increase in countries which haven't vaccinated their population because if people continue to die from COVID, it will put a strain on hospital facilities and people with other illnesses may not get treated. Some children may be orphaned by the deaths of their parents. To prevent the virus spreading too quickly, further lockdowns may be required, which could prevent people from working and they may fall below the poverty line.

**b** It could also impact the economy of richer countries because poorer countries which are struggling might require large sums of aid. Many TNCs which are based in richer countries have factories in poorer countries, so if people are ill or workplaces close down, manufacturing would slow down and profits would fall. If the virus is allowed to spread it may continue to mutate, creating new strains which are resistant to the current vaccines.

**6** Answers will vary.

# 4.9 Escaping from poverty



## About this unit

This unit is about people trying to escape from poverty to create better lives for themselves. It focuses mainly on migrants being smuggled across the Mediterranean from North Africa.

## Key points

- Many people try to escape from poverty and conflict, by moving to other countries.
- Three million migrants, without visas or permits, have crossed the Mediterranean from North Africa to Europe, smuggled by people smugglers.
- There is similar migration across other borders, for example from Mexico to the USA.
- This presents countries with a moral, political, and social dilemma. The richest countries help less than others.

## Key vocabulary

migrant, economic migrant, refugee

## Skills practised in 'Your turn'

- Geographical skills: q2, interpret maps
- Literacy skills: q4, define; q5, rewrite protest messages from placards in your own words
- Thinking skills: q1, q2, q4, explain; q3, suggest reasons; q5, interpret protest messages

## Unit outcomes

By the end of this unit most students should be able to:

- explain the terms given in 'Key vocabulary' above
- explain why so many migrants have crossed the Mediterranean from Africa to Europe
- empathise with people willing to risk everything to find a better life
- describe the dilemma that migrants pose, for other countries

## Ideas for a starter

1. With books closed, say: Close your eyes, and imagine you live in a poor country, with few services. Your family has almost nothing. You have little hope for the future. There are no jobs. You have heard of other places where life is better. What do you think you will do?
2. With books closed, display an image from the Internet of migrants from Africa being rescued in the Mediterranean or the English Channel. (Avoid very distressing images.) Ask the *W* questions: *Who ... What ... Where ... When ...*

## Ideas for plenaries

1. Seydou has no visa or permit for getting into Italy (or any other European country). Suppose he had applied to Italy before he left home, for permission to live there. What do you think the answer would have been?
2. Seydou has to make a speech before a panel, asking for permission to stay in Europe, rather than being sent back to Mali. Students work in pairs or small groups to decide what he will say. Then ask some students to make the speech. (If there is not enough time, this could be completed for homework.)

## Further class and homework opportunities

Suggestions 18–20 on page 107 of this book

*geog.3 workbook*, page 39

**kerboodle**

### Resources

4.9 Skills workshop – Students draw and interpret charts of refugee movement.

4.9 Word scramble – Students complete a crossword of key terms about migration.

4.9 Foundation worksheet – Students recall Seydou's story from pages 82 and 83 of *geog.3* by plotting cities on a map and writing an account of Seydou's journey and future plans.

4.9 Extension worksheet – Students write a letter to their MP outlining the problems faced by migrants trying to reach Europe through Africa.

### Assessment

4.9 End-of-lesson test – An assignable, auto-marked formative assessment.

## Answers to 'Your turn'

**1 a** He wanted to earn money to help his family – who were very poor, with hardly enough to eat. (Mali is one of the world's poorest countries.)

**b** He had to stop three times along the way, to earn money for the next stage of the journey.

**2 a** Zuwara is in Libya. Lampedusa is an island belonging to Italy.

**b** Students will find this one a challenge. From the map they may deduce that Lampedusa is the part of Italy nearest to Libya, and that Libya, being mid-way along North Africa, could draw migrants from many countries, including countries in West Africa, to the east and south of Libya, and Libya itself.

However students will probably not know that migrants reaching the waters around Lampedusa were likely to be rescued by ships from Italy and other EU countries. Or that the breakdown of law in Libya (because of conflict) meant that nobody in Libya was stopping the smugglers. So Libya was an easy exit point.

**3** Money, by preying on the desperation of those who feel they have no option but to migrate. People smuggling is illegal so they know they can charge a lot of money and people will pay it. However, some smugglers may have fallen into the trade due to being in desperate situations themselves.

**4 a i** A person who moves somewhere else, usually to find work, or a better life

**ii** A person who moves in order to find paid work, or earn more

**iii** A person who has been forced to flee from danger, for example from war

**iv** You do not have a document that gives you permission to enter the country

**b** Seydou is a migrant, an economic migrant, and has no permit or visa to enter a European country

**5 a** Answers will vary, but the gist of the message is that it is understandable that people want to make a better life for themselves. We would all do the same.

**b** The gist of the message is that the UK must do something to help prevent migrants from drowning in the Mediterranean. (At some point after this photo was taken, EU countries did agree on a rescue policy, and British ships helped to patrol the Mediterranean and save people. But there was little agreement on what to do about the rescued migrants.)

# 4.10 Can we put an end to poverty?

Help at a glance

## About this unit

This unit is about ways to end poverty: how countries can help themselves, and how other countries, NGOs, and technology, can help too.

## Key points

- Billions of people live in poverty – but it doesn't have to be like this.
- Countries can help themselves by exploiting their resources, cutting out corruption, building hospitals, schools, and infrastructure, and developing manufacturing.
- Development led by a government is called top-down development – but can be expensive.
- Richer countries can help through grants, loans, and expertise, but poorer countries say trade helps more.
- NGOs help through local projects and micro-financing (bottom-up development).
- Technology such as mobile phones can also help, by making it easier to do business, move money, and so on.

## Key vocabulary

top-down development, aid, non-governmental organisation (NGO), microfinancing, bottom-up development

## Skills practised in 'Your turn'

- Literacy skills: q1, summarise within a given word count; q3, define
- Thinking skills: q2, q3, explain; q3, give examples; q3, come up with ideas of how to help; q4, write a reply to an opinion

## Unit outcomes

By the end of this unit most students should be able to:

- explain the terms given in 'Key vocabulary' above
- give one example of top-down development
- give one example of bottom-up development
- explain how the examples given can help to lift people out of poverty
- explain how trade and manufacturing can help to reduce poverty

## Ideas for a starter

1. With books closed, show images from poorer countries all over the world, not just in Africa: for example crowded classrooms with few or no resources, poor hospitals, rural villages without electricity, people carrying water from rivers and firewood, children studying with kerosene lamps. Ask: Can we put an end to poverty?
2. Write on the board: *44% of the world's population lives in poverty.* Then ask questions like: Is it fair? Is it their fault? Should the rest of us do something about it? What can be done?

## Ideas for plenaries

1. You could choose students to read out the text below the images.
2. Use 'Your turn' question 4 as a class discussion.
3. Tell me the three things that made most impact on you, in what you have learned about international development.

## Further class and homework opportunities

Suggestions 21–23 on page 107 of this book

*geog.3 workbook*, page 40

**kerboodle**

### Resources

4.10 Quiz time! – Students select examples of microfinancing projects.

4.10 You decide! – Students rank the poverty reduction strategies.

4.10 Foundation worksheet – Students describe how various people and things, in particular manufacturing, help to reduce poverty in developing countries.

4.10 Extension worksheet – Students explain how certain groups help put an end to extreme poverty and assess whether top-down development is the best option to end world poverty.

### Assessment

4.10 End-of-lesson test – An assignable, auto-marked formative assessment.

## Answers to 'Your turn'

**1** Answers could be along these lines:

- **a** Root out corruption, build infrastructure, educate, train.
- **b** Build factories to employ people, and export.
- **c** Richer countries give aid to poorer countries.
- **d** China gives plenty of aid to Africa.
- **e** Trade fairly with poorer countries.
- **f** Donate to charities working in poorer countries.
- **g** Your donations could provide wells for villages.
- **h** Your donations could help people start businesses.

**2** For example:

- Rooting out corruption means money would go to development projects as intended, rather than into people's pockets. Improved infrastructure, and education and training, all help to give people a chance to earn a living, and move out of poverty.
- Factories give people the opportunity to earn money and hence feed their children better, keep them at school, and so on.
- Trade can help more than aid. By trading fairly, poorer countries will earn more – and that money can help those in poverty.
- A loan to start a business means you can then earn money, feed your children better, keep them at school, and so on.

**3 a i** Top-down development is led by a government. Examples include: exploiting natural resources; building schools and hospitals; educating people; developing the manufacturing sector.

**ii** Bottom-up development is led by NGOs. They help local communities on small scale projects, often by providing training to help them help themselves. Examples include: providing materials for villagers to dig a well; or giving people small loans to help them set up a business.

**b** Students can help with bottom-up development by fundraising. They could organise a bake sale or ask friends and family to sponsor them to do something difficult. Money donated to NGOs such as Oxfam can transform the lives of those in poorer countries.

**4** Answers will vary but are likely to centre around the moral argument that helping those who are less fortunate is simply the right thing to do. There is also a mutual benefit associated with helping poorer countries. For a prosperous world which is environmentally sustainable and politically stable, reducing global inequality is important.

check

## Skills practised in check questions

- Geographical skills: q1, q6, interpret a photo; q1, interpret a maps; q3, interpret graphs
- Literacy skills: q1, q4, use a glossary; q1, q5, define; q1, complete a phrase; q4, comment; q7, extended writing
- Numeracy skills: q3, q4, use HDI scale
- Thinking skills: q1, q4, q5, suggest reasons; q2, interpret a drawing; q3, draw a spider map; q3, q4, q5, explain; q5, imagine a scenario; q6, justify; q6, describe

## Answers for check questions

**1 a** Monrovia

**b** Electricity, gas, running water, sewage pipes, Internet, telephone line. (Utilities are services provided.)

**c** Extreme poverty is where people are living on less than $1.90 a day.

**d** Lack of electricity could hold back development because people would not be able to keep up with the rest of the world. Most industry requires electricity; people may not be able to use computers and the Internet; homes may not have light after dark, so children could not study.

**e** Bottom-up development

**f** 0.4–0.5

**2** GNI per person (PPP)

**3 a** Human development index

**b** Students should draw a spider / sketch map which shows HDI in the centre and each of the three legs should represent: GNI per person (PPP), life expectancy and access to education.

**c** Armenia has a higher HDI

**d** It is likely that Armenia gives children a better start in life as HDI includes the number of years schooling that children can expect. However, since this measure of development also incorporates life expectancy and the number of years adults have been schooled for, students cannot confirm this statement with certainty.

**e** There is extreme inequality in Equatorial Guinea. Students can deduce this from the graphs. It has the same GNI as Armenia, yet the HDI score is much lower. This indicates that there are many social issues in the country. Low life expectancy could point to inadequate health care, malnutrition, lack of sanitation, or loss of young lives in conflict. A low score for access to education could mean that adults only received a few years of schooling; and / or children are not adequately educated.

**f** HDI gives a truer picture because it includes both social and economic measures.

**4 a** The trend shows increasing HDI in all four countries.

**b** This is a good thing because it indicates development and progress. People are earning more money and there are improvements in health and education.

**c** This is because it developed its industrial sector very quickly. Fewer people live by farming, and many now live in cities, earning higher wages than those in rural areas. In cities, people have better access to services such as hospitals and schools.

**d** Emerging economies are those which are developing quickly, largely as a result of industry.

**e i** One from: Malawi is landlocked; the climate is hot and dry; the north is very mountainous; it has very few natural resources.

**ii** One from: Malawi was colonised, which left it very poor; most of its employment is in the primary sector which brings in little income; most of its exports are raw materials; it has few mineral resources; there is corruption in the government.

**f** The 2002 famine caused thousands of deaths. Having no food to sell lowered Malawi's income. People were not able to make personal progress due to hunger. If a person's main goal is to feed themselves, they have no time for other activities which may aid a country's development, such as participating in education.

**g** The UK's HDI is the highest in graph D. This shows it is an advanced, high income country. A score above 0.8 indicates it has a high life expectancy, adults had good access to education (many years in schooling), and children can expect the same. Overall, there is a good standard of living in the UK and a high GNI per person.

**5 a** A transnational corporation (TNC) is a company which has branches and operations in more than one country.

**b** For a Chinese TNC making cotton clothing and towels, Malawi may have been chosen because: cotton is grown there so there will be good access to raw materials; it has a high fertility rate so there will be lots of young people looking for work; many people still live by farming but may wish to earn higher wages doing factory work; Malawi is poorly developed so the TNC will be able to pay low wages, which means more profit.

**c** Malawi's president would be happy about this because: TNCs can increase GNI; taxes can be used to build more hospitals and schools; the TNCs may build new infrastructure such as roads and railways; new technology may be introduced; if people earn higher wages they can improve their lives.

**6** Answers might include: conflict / war, pandemics / disease, climate change, natural disasters.

**7** Good answers will give evidence to support both sides of the argument and provide a short conclusion.

An example answer: There are many ways richer countries can help poorer ones to develop. Aid and / or cheap loans are often part of this picture, and expertise may be provided for big projects like roads. This help often comes from TNCs rather than directly from governments. Trade is also important because if richer countries buy products from poorer countries, it allows them to make a profit. NGOs run development projects which can transform lives from the bottom up.

On the other hand, poorer countries can also help themselves, by rooting out corruption from their governments and by developing their manufacturing and tourism sectors.

To some extent, poorer countries do need help from richer ones, but the best outcomes occur when they combine any help with their own efforts.

# Further suggestions for class and homework

Most of these suggestions are addressed to your students. Where research or further resources are needed, the Internet will almost certainly provide the answer.

The suggestions are graded *, **, *** according to level of difficulty. Some are suitable for all levels, and can be differentiated by outcome.

## An unequal world

1 **Create a presentation** Create a presentation on the theme *An unequal world*. You could use PowerPoint, video, or audio. Grab people's attention, and make them think. ***/**/*****

2 **Measuring poverty** How else could you measure poverty? Is it just about how much money people have? Think up ways to measure it – without looking in books. ***/**/*****

## So ... what is development?

3 **Development: a process of change that improves people's lives** It can go on anywhere, at any level. Is it going on in the UK? Look in the papers and watch the news for a week, to find out. What issues are people talking about? Which do you think are about development? Tell the class. *******

4 **Living in a welfare state** The UK is a welfare state. So even if you are poor you can still get health care, education, a place to live, and some money to live on. Find out more about the welfare state and do a five-minute presentation to the class. How will you structure it? ****/*****

## How is development measured?

5 **Make up some development indicators** See how many (sensible) development indicators you can come up with, without looking in books. How about happiness? ****/*****

6 **Map life expectancy** A challenge for the determined. Create your own choropleth map for life expectancy (or another indicator) on an outline world political map. Find the data on the CIA World Factbook website. Which ranges will you use? What patterns does your map show? ****/*****

## How is Malawi doing?

7 **Malawi display** Create a class display about Malawi. Brainstorm the content (map? flag? people? culture? climate? wildlife? a time line for its history? photos? recipes? phrases in Chichewa, its main language?). The work could be divided among individuals or groups, and run for the duration of the international development topic. ***/**/*****

8 **A holiday in Malawi** Use the Internet to find out what Malawi offers tourists. Then plan a 10-day holiday there. Prepare an itinerary, with a map. Include time, date, and mode of departure from the UK. (Will you fly? From where? To where?) Will you need any vaccinations? ***/**/*****

## How did the development gap grow? – part 1

9 **Why is the UK so developed?** Present reasons, in any form you choose, to explain why the UK is one of the world's more developed countries. ****/*****

10 **Is the UK becoming more developed?** What do you think? How could you find out if you are right? (Indicators?) *******

11 **Development and natural hazards** Bangladesh suffers flooding most years. Find out more, and draw a consequence map to show how this could hinder development. ****/*****

## How did the development gap grow? – part 2

12 **Why does Malawi lag behind in development? – an enquiry** Carry out an enquiry to find out why Malawi is poorly developed. Don't forget the different kinds of reasons: historical, geographical, and social (health, education, and so on), exports, industry, etc. Note: this is challenging! *******

13 **You as entrepreneur** You are a bright young Malawian. Malawi grows cotton and exports the raw fibre. Come up with an idea to process the fibre and add value, for example by making clothing. Give as much detail as you can. Make up a name for your company and design a logo. ***/****

## Globalisation, development, and TNCs

14 **More about BP** What kind of business is it? Where are its headquarters? Where in the world does it operate? What else can you find out? Tell us – in any way you like. ***/**/*****

15 **Acrostic** Use the word GLOBALISATION to create an acrostic about globalisation, development, and TNCs. ****/*****

## COVID-19 and vaccine inequality

16 **COVID now** What's happened since 2019? What is the world like now in terms of COVID? Is the world fully vaccinated? Create a poster / wall chart. ****/*****

17 **Imagine this!** A new pandemic has emerged. What should we do this time to protect people? The same as for COVID, or different? You are the prime minister. You decide! ****/*****

## Escaping from poverty

18 **Explore Lampedusa** Find out more about this Italian island. (You could look at it on Google Earth. ) Write an entry about it for a geography book about Italy. ****/*****

19 **Seydou's story** It had a happy ending. Tell us! ***/**/*****

20 **Crossing the Mexican border** Write a story about a young Mexican, trying to cross from Mexico into the USA without a permit. Think about *Who ... What ... Where ... Why ... When ...* ***/**/*****

## Can we put an end to poverty?

21 **What if ...** there was no more poverty in the world? Respond to this question in any form you choose. ***/**/*****

22 **Fair Trade ...** The Fair Trade movement is trying to end poverty. Find out how it works, and tell the class. Give examples of Fair Trade goods in stores near you. ****/*****

23 **Do we really need more and more?** Do we need to keep buying more and more things, and using up Earth's resources? Should we focus on helping to eradicate poverty instead, and make the world more equal for everyone? Find a suitable way to respond. ****/*****

# 5 Our restless planet

## About this chapter

### Addressing the KS3 Programme of Study

- Understand, through the use of detailed place-based exemplars at a variety of scales, the key processes in physical geography relating to … plate tectonics
- Extend locational knowledge and deepen spatial awareness of the world's countries

### About the content

- The 2008 Sichuan earthquake is included as a case study for several reasons: Sichuan is in China, studied in *geog.2*. The earthquake is the costliest in China in financial terms; and there were many deaths (87 000, including many students).
- The 2004 tsunami is included because it was the most devastating tsunami known. The case study also gives an opportunity to add to locational knowledge around the Indian Ocean.
- Mount Vesuvius was chosen because it is one of the word's most dangerous volcanoes, and is located in a densely populated area.

## The big picture

These are the key ideas in this chapter.

- Earth consists of three layers: the crust, mantle, and core.
- The hard, rigid, outer part of Earth is cracked into big slabs, which we call plates. These move around.
- Plates can move apart, push into each other, or slide past each other. The movements cause earthquakes – and some cause volcanic eruptions – at or near the plate edges.
- Earthquakes and volcanic eruptions cause a great deal of damage, and loss of life.
- So people living around or near plate edges are at risk from earthquakes and eruptions – but they do not usually move away.

## The chapter outline

Use this to give students a mental roadmap for the chapter.

5 **Our restless planet** The chapter-opening photo in the students' book is an important resource. See opposite for using it.

5.1 **What's Earth like inside?** The Earth's layers and a closer look at the crust and mantle

5.2 **A cracked Earth?** Plates, and their link with earthquakes and volcanoes

5.3 **A closer look at plate movements** Different kinds of plate movements, how they cause earthquakes and eruptions, and mountain building

5.4 **What is an earthquake?** What they are, the Richter scale, and the damage earthquakes do

5.5 **An earthquake in Southwest China** Case study: the major earthquake that killed over 87 000 people in Sichuan in 2008

5.6 **Tsunami!** Case study: the devastating tsunami in the Indian Ocean in 2004, which affected 15 countries

5.7 **What is a volcano?** The structure of volcanoes, and the damage eruptions can do

5.8 **The most dangerous volcano in the world?** Case study: Mount Vesuvius; why it is dangerous; evacuation plans

5.9 **So why live in a danger zone?** Why people continue to live in places along or near, plate margins, when there's a risk of earthquakes and eruptions

5 **Our restless planet check questions** A page of questions to check knowledge, understanding, and skills across the chapter, with an opportunity for extended writing

## Objectives and outcomes for this chapter

| Objectives | Unit | Outcomes |
|---|---|---|
| Most students will understand: | | Most students will be able to: |
| • that Earth is made up of different layers | 5.1 | • name and describe the three layers that make up Earth |
| • that Earth's hard outer part (the lithosphere) is cracked into big slabs called plates, which move | 5.2 | • explain what a plate is and why it moves; name at least five plates; explain the pattern of earthquakes and volcanoes on a world map |
| • that plate movements cause earthquakes, and some lead to volcanic eruptions, and cause mountain building | 5.3 | • explain that earthquakes and volcanoes tend to occur along or close to plate margins; describe the three ways in which plates move relative to each other; explain why each causes earthquakes, and some lead to eruptions, and fold mountains |
| • what earthquakes are, what kind of damage they do, and why | 5.4 | • explain what causes earthquakes, and why everything shakes; give examples of earthquake damage; explain what the Richter scale tells us |
| • why south-west China is prone to earthquakes | 5.5 | • explain why the 2008 Sichuan earthquake occurred; describe the damage it did |
| • what tsunamis are, and what kind of damage they do | 5.6 | • explain what causes a tsunami, and why it can affect many countries; describe the damage it can do |
| • what volcanoes are, and what kind of damage eruptions cause | 5.7 | • draw a labelled cross-section of a volcano; list the products from eruptions, and the damage they cause |
| • why Mount Vesuvius is such a dangerous volcano | 5.8 | • say where Mount Vesuvius is; give two reasons why it is dangerous |
| • why millions of people live along, and near, plate edges, even though they are at risk from tectonic activity | 5.9 | • give at least four reasons why people continue to live in tectonic danger zones |

These tie in with the outcomes for each unit in this teacher's handbook.

## Assessment for this chapter

There are assessment materials for this chapter on *geog.3 Kerboodle Lessons, Resources, & Assessment*. They include interactive auto-marked end-of-lesson and chapter-level tests, a printable exam-style end-of-chapter paper (with a mark scheme), and a printable self-assessment checklist. See also the notes on pages 14–17 of this book.

## Getting ready for this chapter

There is a scheme of work for this chapter, plus a glossary and glossary worksheet, on *geog.3 Kerboodle Lessons, Resources, & Assessment.*

## Using the chapter-opening photo

Ask students:

What's that black stuff flowing across the road?

Where did it come from?

It's hot. What do you think it will be like when it cools?

The photo shows lava flowing across a road on Big Island, Hawaii.

This island is in the middle of the Pacific plate. So its volcanic activity is unusual in that it is not due to tectonic activity at plate margins. Rather, it is at a volcanic 'hotspot' where molten magma breaks through a plate. If the hotspot is under the ocean, underwater volcanoes form. Eventually they poke up as islands.

As the Pacific plate slowly moved north-west over the Hawaiian hotspot, over millions of years, the line of volcanic islands which we call Hawaii were formed.

# What's Earth like inside?



## About this unit

This unit is about Earth's structure. It shows the layers which make up Earth, and then focuses on the crust and mantle.

## Key points

- Earth is made up of three different layers: crust, mantle, and core.
- Earth gets hotter as you go deeper.
- The crust (which we live on) is solid rock; rock in the upper mantle is hard and rigid, in the lower mantle it is solid, but not rigid; the core is a mixture of metals – iron and nickel; the inner core is solid, the outer core is liquid.
- There are two types of crust: continental crust which forms the continents, and is mainly granite; and oceanic crust which lies under the oceans, and is mainly basalt.
- The crust and upper mantle form a hard layer called the lithosphere. This floats on top of the lower mantle.

## Key vocabulary

crust, mantle (upper and lower), core (inner and outer), oceanic crust, continental crust, granite, basalt, trench, lithosphere, convection current

## Skills practised in 'Your turn'

- Geographical skills: q6: drawing and labelling a cross section
- Literacy skills: q1, fill in a table
- Numeracy skills: q2, addition and division
- Thinking skills: q3, q5, explain; q3, compare; q4, describe

## Unit outcomes

By the end of this unit most students should be able to:

- explain the terms given in 'Key vocabulary' above
- name and describe the three layers which make up Earth
- explain the difference between continental and oceanic crust
- draw a simple diagram to show continental and oceanic crust, the mantle, and the lithosphere

## Ideas for a starter

1. With books closed, ask: What do you think Earth is like inside? Write suggestions on the board. Say you'll come back to them later (see plenary 3).
2. Ask students to imagine they are digging a tunnel down through Earth, under the school. What do they think it will be like at a depth of: 1m? 10m? 100m? 1km? 1000km? 6000km? Ask them to close their eyes and do a mind movie.

## Ideas for plenaries

1. The deepest hole drilled so far is only just over 12 km. Name a place 12 km from here. So how do we know so much about inside Earth? (By studying earthquake waves, and meteors from other planets.)
2. Ask students to each write an answer, on a strip of paper, about the work in the unit. For example: *It's mainly liquid iron.* Pool the answers. The class then guesses the questions.
3. Return to students' ideas on the board, from starter 1. Which can be ticked? Which should be crossed out or corrected? What else can be added?

## Further class and homework opportunities

Suggestions 1–3 on page 129 of this book

*geog.3 workbook*, page 41

kerboodle

### Resources

5.1 Sort it out! – Students label a diagram of Earth's layers.

5.1 Quiz time! – Students complete a table to describe Earth's layers.

5.1 Foundation worksheet – Students complete a table on the layers that make up the Earth and label a diagram of the Earth's crust.

5.1 Extension worksheet – Students write a first-person account imagining they have taken a journey to the centre of the Earth.

### Assessment

5.1 End-of-lesson test – An assignable, auto-marked formative assessment.

## Answers to 'Your turn'

1

| Layer | Made of ... | Solid or liquid? | How thick? |
|---|---|---|---|
| Crust | rock | solid | 7–70 km |
| Mantle – upper | rock | solid | 600 km |
| Mantle – lower | rock | solid but not rigid | 2300 km |
| Core – outer | iron and nickel | liquid | 2200 km |
| Core – inner | iron and nickel | solid | 1260 km |

2 a About 6400 km.

b 322.5 hours. If you cycle eight hours a day that's just over 40 days. Pretty strenuous cycling!

3 a Continental crust is mainly granite, which is an igneous rock. It forms the land.

b Oceanic crust is made of a heavier igneous rock, basalt. It is much thinner than continental crust and is covered by sediment.

4 The lithosphere consists of Earth's crust and its upper mantle. It is a layer of rigid rock which floats on top of the lower mantle.

5 In the lower mantle, the rock is solid but not rigid. This means it can flow very slowly.

6 The missing labels, from left to right, are:

ocean

oceanic crust made of basalt

upper mantle

continental crust made of granite

the crust and upper mantle together form the lithosphere.

# A cracked Earth?

Help at a glance

## About this unit

This unit looks at how the pattern of earthquakes and volcano sites are linked to Earth's plates.

## Key points

- Earthquakes are caused by sudden rock movements.
- Volcanoes form where liquid rock shoots out at Earth's surface.
- The lithosphere is cracked into huge slabs called plates, which move.
- Earth's main earthquake and volcano sites lie along, or near, plate boundaries (margins).
- Plates move as a result of gravity (slab pull and ridge push).

## Key vocabulary

earthquake, volcano, plate, Ring of Fire, lithosphere, plate boundary (margin), gravity, slab pull, ridge push

## Skills practised in 'Your turn'

- Geographical skills: q2, q3, q4, q6, interpret maps; q2, q3, compare maps
- Literacy skills: q1, q5, define
- Thinking skills: q3, explain patterns; q4, suggest reasons; q6, explain processes

## Unit outcomes

By the end of this unit most students should be able to:

- define the terms given in 'Key vocabulary' above
- compare maps to explain the pattern of earthquake and volcano sites around the world
- explain what Earth's plates are, and name at least five of them
- explain how plates move

## Ideas for a starter

1. Write the headings *Earthquakes* and *Volcanoes*, one on each half of the board. Then ask: What do you know about earthquakes? What do you know about volcanoes? Write down words or phrases to summarise students' responses. Could they be grouped in any way? For example: *causes / consequences / locations?*
2. Hold up an egg (hard-boiled, but don't say). Ask: What has this and Earth got in common? (Hard outer part, layers inside.) Then tap on the shell and crack it. Ask: Now what else do it and Earth have in common? (Hard outer part cracked.)

## Ideas for plenaries

1. Display Earth on *Google Earth*, or a relief map of Earth, including the ocean floor. Students try to identify plate edges, by comparing it with the map on page 91 of their books.
2. Tell me the three key facts you learned today. Students discuss and agree on them. (For example: there are plates; they are moving; the movement gives rise to earthquakes and volcanoes.)
3. Choose a student to be in the hot seat. With books closed, another student asks them a question about what they've learned in today's lesson. Then nominate two different students (4–6 pairs in total). There's one golden rule: questions must not be repeated!

## Further class and homework opportunities

Suggestions 4 and 5 on page 129 of this book

*geog.3 workbook*, page 42

**kerboodle**

### Resources

5.2 Word scramble – Students complete a crossword of key terms about tectonic plates.

5.2 Quiz time! – Students complete a quick quiz about Earth's tectonic plates.

5.2 Foundation worksheet – Students fill in the missing words in a paragraph describing the lithosphere and identify true statements about plate movement.

5.2 Extension worksheet – Students fill in the missing words in a paragraph describing the lithosphere and label a diagram explaining how plates move.

### Assessment

5.2 End-of-lesson test – An assignable, auto-marked formative assessment.

## Answers to 'Your turn'

**1** **a** A shaking of Earth's crust. It is caused by a sudden rock movement.

**b** A volcano is where lava (melted rock) erupts out through Earth's surface.

**c** A plate is part of Earth's hard outer crust, which is broken into slabs – called plates.

**2** **a** Eurasian **b** North American

**c** Indo-Australian or African **d** Pacific

**e** Nazca or Antarctic **f** Pacific

**3** Both maps show earthquake sites and volcanoes. Map C also shows Earth's plates and the direction they are moving. The location of the plates explains the pattern of earthquakes and volcanoes. Most of these hazards are located at plate boundaries because plate movement causes them. However, there are also a few hazards not located on plate boundaries, such as the earthquakes on the Eurasian plate, north of India.

**4** **a** The UK has no active volcanoes today because it is not near a plate boundary.

**b** The UK has the remains of several ancient volcanoes because it used to be on a plate boundary. Examples are: Edinburgh, which sits on the remains of an extinct volcano; and the mountains in north Wales which are the remains of a volcanic plateau.

**5** A plate margin is the region where two plates meet; also known as a plate boundary.

**6** **a** Nazca and South American plate

**b** **i** Nazca

**ii** Oceanic plates are heavier than continental plates, so where the Nazca plate meets the coast of South America, it sinks.

**iii** The plate is kept moving by the process of slab pull, which is the weight of gravity pulling it down into the mantle.

**iv** At the opposite edge of the plate, the plates are being pushed apart at the margin because the South American and the African plate are moving in different directions. The process of ridge push is caused by rock melting and forming new land between the plates.

# 5.3 A closer look at plate movements

Help at a glance

## About this unit

In this unit students explore different types of plate movement, and how they can lead to earthquakes, volcanic eruptions and mountain building.

## Key points

- Constructive plate margins occur where plates are moving apart (diverging). Volcanoes and earthquakes occur on these plate margins.
- Destructive plate margins occur where plates push into each other (converging). There are earthquakes and volcanoes on these plate margins.
- Conservative plate margins occur where plates slide past each other. Earthquakes occur on these plate margins.
- Where plates push into each other fold mountains may be created, eg the Himalayas and the Andes.

## Key vocabulary

diverging, magma, volcano, constructive plate margin, converging, ocean trench, destructive plate margin, conservative plate margin, fold mountains

## Skills practised in 'Your turn'

- Geographical skills: q2, draw and label a diagram; q3, interpret maps
- Literacy skills: q1, complete a table, and a sentence; unjumble a word; q3, define
- Thinking skills: q3, q4, q5, explain

## Unit outcomes

By the end of this unit most students should be able to:

- explain the terms given in 'Key vocabulary' above
- describe the three ways in which plates move relative to each other
- explain that all plate movements cause earthquakes
- say which two types of plate movement give rise to volcanoes and why
- explain how fold mountains are formed, and name two ranges of fold mountains

## Ideas for a starter

1. With books closed, ask quick-fire revision questions. What are plates? Name some. Why do plates move? What is slab pull? Ridge push?
2. Show students image D on the board. Make it as large as possible. Ask: What is that pale curved line running down the Atlantic Ocean? What is that dark line running part way along the western coast of South America?

## Ideas for plenaries

1. Discuss the answers to 'Your turn' question 1. Make sure students understand the reasons behind the ticks and crosses. (They are in the table on the next page.)
2. Do you like the idea that plates are moving? And that you are being carried around on one? Why? / Why not?

## Further class and homework opportunities

Suggestions 6 and 7 on page 129 of this book

*geog.3 workbook*, page 43

kerboodle

### Resources

5.3 Quiz time! – Students complete a quick quiz to identify the effects of different types of plate movement.

5.3 Sort it out! – Students complete a table to describe the effects of plate movements.

5.3 Foundation worksheet – Students label how plates move, identify areas where earthquakes are likely to happen, and order the process for volcanoes forming.

5.3 Extension worksheet – Students use pages 92 and 93 of *geog.3* to fill in the gaps of a paragraph explaining plate movements.

### Assessment

5.3 End-of-lesson test – An assignable, auto-marked formative assessment.

## Answers to 'Your turn'

**1 a** Explanatory notes have been added to this table for you.

| Plate movement ... | ... gives earthquakes | ... gives volcanoes | ... builds mountains |
|---|---|---|---|
| 1 Pushing into each other | ✔ all plate movements give earthquakes | ✔ wherever oceanic crust is pushed down into the mantle | ✔ wherever rock is pushed and folded upwards |
| 2 Sliding past each other | ✔ | ✗ no rock melts | ✗ no rock is pushed upwards |
| 3 Pulling apart | ✔ | ✔ magma wells up between them | ✔ volcanoes may grow into underwater mountains |

**b** Earthquakes

**c** Converge

**2** The drawing should look like the one below. It should have the title, 'Destructive plate margin.' Note that earthquakes will occur anywhere that rock gives way under pressure.

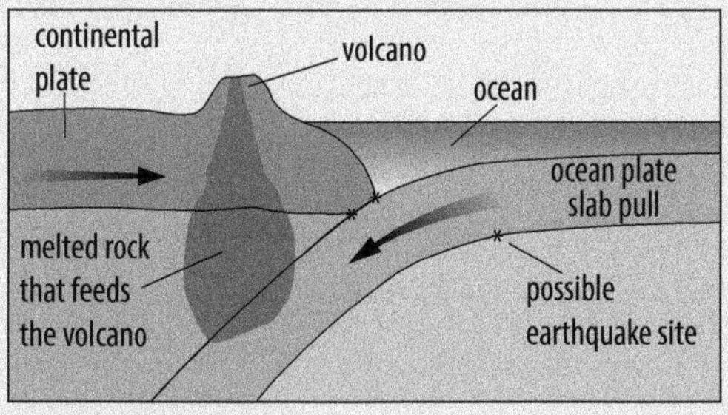


**3 a** Mountains formed by plates pushing into each other; rock at the plate edges gets folded upwards, making mountains

**b i** Peru is on the South American plate (continental crust), near the boundary with the Nazca plate (oceanic crust). The two plates are pushing into each other. The denser Nazca plate is diving downwards, causing earthquakes and volcanic eruptions in Peru. Rock is getting squeezed upwards too, giving fold mountains: the Andes.

**ii** Italy is on the Eurasian plate, but very close to the boundary with two other plates. All three plates are pushing into each other. Heavier oceanic crust is diving under the continental crust, causing earthquakes and volcanoes. Land has been folded upwards too, giving Italy's Apennine Mountains.

**4 a** Destructive – as plates push into each other, land is 'destroyed'. When an oceanic plate meets a continental plate, the plate under the ocean floor sinks into the mantle, where it melts and is destroyed in the process. If two oceanic plates meet, whichever plate is the heaviest will sink and melt. If two continental plates push together, land is destroyed by the crumpling and folding process.

**b** Conservative – when plates slide past each other, no lithosphere is destroyed – hence it is 'conserved.'

**c** Constructive – when plates move apart, rock melts and rises, 'constructing' new land.

**5** Although the ocean floor is getting wider in places, it is being destroyed in other places, where it is diving under continental crust.

# What is an earthquake?



## About this unit

This unit explains what earthquakes are, introduces the Richter scale, and summarises the damage that earthquakes can do.

## Key points

- Earthquakes are caused by sudden rock movement, when rock under strain gives way.
- The strain energy that has been building up in the rock is suddenly released, and spreads in all directions as seismic waves.
- These waves shake Earth as they travel through it. Structures standing on Earth's surface get shaken in turn, and may collapse.
- The magnitude of an earthquake is the amount of energy it releases. It is measured on the Richter scale.
- An increase of 1 on the Richter scale represents a 10-fold increase in the amplitude of the waves, and around a 30-fold increase in released energy – meaning a lot more damage.

## Key vocabulary

earthquake, strain energy, seismic wave, focus, epicentre, aftershocks, seismometer, magnitude, Richter scale, tsunami

## Skills practised in 'Your turn'

- Geography skills: q3, interpret a photo: q4, interpret a diagram; q5, interpret maps
- Numeracy skills: q4 requires students to understand the Richter scale
- Literacy skills: q3, describe in a given number of words
- Thinking skills: q2, q4, q5, explain

## Unit outcomes

By the end of this unit most students should be able to:

- explain the terms given in 'Key vocabulary' above
- explain what causes earthquakes, and why everything shakes
- explain what the Richter scale is, and that the higher the Richter number, the more damage an earthquake can do
- give examples of earthquake damage

## Ideas for a starter

1. Cover a heavy brick with sandpaper, taped in place. Loop a length of elastic around the sides. Place it on a second sheet of sandpaper and pull on the elastic. Watch what happens as the tension mounts. (The stretch in the elastic indicates 'strain'.)
2. With books closed, display any image from the Internet of earthquake damage in a city. Ask: What do you think happened here?

## Ideas for plenaries

Plan plenaries for strategic points throughout the lesson, as well as at the end.

1. Are there earthquakes in the UK? Answer: Yes, a few hundred every year, but only 20–30 are large enough for people to notice.
2. Call out up to 10 statements about earthquakes, some true and some false. For example: Earthquakes occur only along plate edges. The epicentre is directly below the focus. Students have to raise green or red cards, or shout out *True* or *False*. Where a statement is false, nominate a student to correct it.
3. Search the Internet for *Where have there been earthquakes today?* Display the map, and discuss their locations and magnitude.

## Further class and homework opportunities

Suggestions 8 and 9 on page 129 of this book

*geog.3 workbook*, page 44

**kerboodle**

### Resources

5.4 Sort it out! – Students sort the causes and effects of earthquakes.

5.4 Quiz time! – Students interpret a diagram of an earthquake.

5.4 Foundation worksheet – Students assign pictures and descriptions to numbers on the Richter scale, explaining what an earthquake at each value would look like.

5.4 Extension worksheet – Students categorise the primary and secondary impacts of earthquakes, and consider how P and S waves can be used to protect people.

### Assessment

5.4 End-of-lesson test – An assignable, auto-marked formative assessment.

## Answers to 'Your turn'

**1** When enough tension builds up in a rock it will suddenly give way, like an elastic band snapping. The result is an earthquake.

**2 a** Seismic waves are waves of energy given out during an earthquake; they pass through Earth in all directions and cause everything to shake.

**b** The focus of an earthquake is the point where the rock moved, setting the earthquake off.

**c** The epicentre of an earthquake is the point on Earth's surface directly above the focus.

**d** The magnitude of an earthquake means how much energy it gives out; it is measured on the Richter scale.

**3** Answers will vary. Encourage students to look closely at the photo.

**4 a** At A, since it is nearer the epicentre.

**b** At B, since buildings can collapse and kill people.

**c** A lot more damage, since it releases a lot more energy.

**d** 30 times

**e** At 5 am, since people will be asleep. By 10.30 am, many will have left for work and school. That does not mean they are out of danger. But they will have more chance to protect themselves if they are awake, for example by moving away from windows, and following the emergency earthquake drill.

**5 a** Chile is near the edge of the South American plate, close to where the heavier Nazca plate is sliding under it. So Chile has volcanoes, and many earthquakes.

**b** Chile is a long, narrow country, and most of its towns and cities are close to the coast, so they were badly affected by this large earthquake which occurred at the plate edge. The earthquake also set off a tsunami, which caused death and destruction along the coast. At the time, Chile was still a relatively poor country, and few buildings were built to withstand earthquakes. Many were poorly constructed, so they collapsed and killed people.

**c** The earthquake set off a tsunami in the Pacific Ocean which travelled across the ocean, and drowned people in Japan. (It also drowned people in Hawaii and did extensive damage in Hawaii and Japan, and some damage on the West Coast of the USA.)

# An earthquake in Southwest China



## About this unit

This is a case study of the 2008 Sichuan earthquake, in which over 87 000 people died. The dead included thousands of students killed when badly-built schools collapsed.

## Key points

- Sichuan is in Southwest China.
- Southwest China is prone to earthquakes, because it is near the boundary where the Indo-Australian plate is pushing into the Eurasian plate.
- Earthquakes don't kill. Collapsing structures do.
- Buildings collapsed in the Sichuan earthquake, because they were not built to cope with shaking. Over 87 000 people died. And over 370 000 people were injured.
- A shake map shows the intensity of shaking across the region where an earthquake occurs.

## Key vocabulary

earthquake, shake map, fault, Indo-Australian plate, Eurasian plate, epicentre

## Skills practised in 'Your turn'

- Geography skills: q1, give location; q2, interpret a photo; q4, q5, interpret a shake map
- Numeracy skills: q5, use a scale
- Literacy skills: q2, describe an imaginary experience; q3, write an explanation in a given number of words
- Thinking skills: q3, q5–q7, explain

## Unit outcomes

By the end of this unit most students should be able to:

- explain the terms given in 'Key vocabulary' above
- say where Sichuan is
- explain what caused the earthquake
- explain why earthquakes are deadly

## Ideas for a starter

**1** If you did not do starter 2 on page 116, with books closed, display an image from the 2008 Sichuan earthquake. (Avoid the very distressing ones.) Ask *Who … What … Where … Why …* questions. Say that the class will now find out more about this earthquake.

**2** With books closed, display a set of jumbled words about earthquakes. For example:

| | | | | | |
|---|---|---|---|---|---|
| *ashek* | *paesollc* | *lilk* | *scouf* | *ercenpite* | *aginteudm* |
| (*shake* | *collapse* | *kill* | *focus* | *epicentre* | *magnitude*) |

Give students three minutes to unjumble them.

## Ideas for plenaries

**1** You could choose a student or students to read out the account of the earthquake.

**2** After question 5, discuss the shake map further. For example, why is the severe shaking along a rectangle, rather than in a circle? (Because rock slipped downwards along a long fault.)

**3** Write on the board: *Earthquakes do not kill. Collapsing buildings do*. Then ask for suggestions for the design of a building which can withstand earthquakes.

## Further class and homework opportunities

Suggestions 10 and 11 on page 129 of this book

*geog.3 workbook*, page 45

**kerboodle**

### Resources

5.5 Case study: The Sichuan Earthquake – Students complete a case study about the Sichuan Earthquake.

5.5 Quiz time! – Students complete a quick quiz about the Sichuan Earthquake.

5.5 Foundation worksheet – Students illustrate and label their day as a student in Beichuan Middle School in Southwest China during the 2008 earthquake.

5.5 Extension worksheet –Students write a diary entry imagining they are a student in Beichuan Middle School in Southwest China during the 2008 earthquake.

### Assessment

5.5 End-of-lesson test – An assignable, auto-marked formative assessment.

## Answers to 'Your turn'

1 It is in Asia, in the southwest of China.

2 Answers will vary.

3 The answers should include the main points from panels 1–4 at the top of page 97.

4 **a** Map C is called a shake map because the choropleth scale represents the intensity of the ground motion and the shaking experienced during the earthquake.

**b** The wrinkles show that the land is hilly or mountainous. (The text describes it as mountainous. The diagram shows hills.)

5 **a** The red represents the area with the most severe shaking.

**b** They are both the same distance from the epicentre, but Mianyang is closer to the fault line along which the rock slipped downwards.

**c** About 250 km.

6 There will be more earthquakes, because the plates are still moving and pushing against each other.

7 Most deaths will have been caused by buildings collapsing. (Some may have been caused by fires, or by roads collapsing.) If the buildings had been stronger and earthquake-proofed most deaths could have been prevented.

# 5.6 Tsunami!



## About this unit

This unit is about the devastating tsunami in the Indian Ocean in December 2004. It explains what a tsunami is, and describes the widespread damage caused by this tsunami.

## Key points

- A tsunami is a series of waves set off in the ocean, by an earthquake in the ocean floor. The waves travel through the water in all directions.
- Out in the ocean they travel fast, but are not high. As they approach shallow water they slow down, but grow taller – sometimes as high as 30 metres.
- When they hit land these powerful waves can do enormous damage.
- The Indian Ocean tsunami killed over 227 000 people.

## Key vocabulary

tsunami, earthquake, epicentre

## Skills practised in 'Your turn'

- Geography skills: q2, q4, interpret maps
- Thinking skills: q1–q3, q5, explain; q4, give reasons

## Unit outcomes

By the end of this unit most students should be able to:

- explain the terms given in 'Key vocabulary' above
- explain what a tsunami is, and what causes it
- explain why it can affect many countries
- describe the damage it can do

## Ideas for a starter

1. With books closed, ask: Suppose there was a massive earthquake in the ocean floor. What do you think it would do to the water?
2. Ask: Can anyone tell me what a tsunami is? What causes it? Can you tell me anything else about it? Students write answers on the board, perhaps in the shape of a high wave, or around a drawing of a high wave.
3. With books closed, display a photo of tsunami damage, or a tsunami, from the Internet (it could be any tsunami). For example, a photo showing a large boat on top of a building would be a good prompt. Ask *Who … What … Where … Why …* questions about it. You could write students' answers on the board, and say you'll come back to them later.

## Ideas for plenaries

1. At the start of class, give out slips of paper numbered 1 to 12, one to each of 12 students. Then when you reach the maps on page 96 of the students' book, they read out the text in the corresponding boxes. Ask them to say *15 minutes after the earthquake* and *two hours after the earthquake* instead of *+15 minutes*, *+2 hours,* and so on.
2. Ask for comments about the photos at the top of page 97 of the students' book.
3. If you chose starter 2 or 3, you could return to it. Ask students to modify, or add, to what was written on the board.
4. So … you are on a beach. There is no tsunami warning system. How might you know that a tsunami is approaching? What will you do?
5. Show an animation of a tsunami.

## Further class and homework opportunities

Suggestions 12 and 13 on page 129 of this book

*geog.3 workbook*, page 46

**kerboodle**

### Resources

5.6 You decide! – Students rank the reasons why so many people died in the 2004 tsunami.

5.6 Quiz time! – Students complete the explanation of the Indian Ocean tsunami.

5.6 Foundation worksheet – Students identify the differences between two photos, one taken before a tsunami and one taken after.

5.6 Extension worksheet – Students write a script for a radio interview, explaining two common assistance methods and why they are important to people recovering from disaster.

### Assessment

5.6 End-of-lesson test – An assignable, auto-marked formative assessment.

## Answers to 'Your turn'

**1** It's caused by an earthquake in the ocean floor (and the earthquake is caused by plate movements).

**2** **a** Many countries border the Indian Ocean. The waves travelled across the ocean in all directions, and reached over a dozen of them. (Some countries were safer because they were sheltered by others.)

**b** Some countries were further from the epicentre than others, so it took longer for the waves to reach them.

**c** The epicentre was very close to Indonesia. Somalia was thousands of kilometres away. The waves lost energy as they travelled, so they did less damage in Somalia.

**d** The waves may be only a metre high out at sea, and in many ocean-going boats you would not notice this. The waves slow down and grow taller in the shallower water at the coast.

**e** The Philippines were sheltered from the tsunami by Thailand and Indonesia. (Students should check the map on pages 140–141 of the students' book.)

**3** Earthquakes shake buildings and other structures, and people are killed when these collapse. Tsunami waves rush onto land with great force, and can wash away buildings, and also drown people far from buildings, and wash away crops, farm animals, and other forms of livelihood. There are some things you can do to protect yourself during an earthquake. There is little you can do to protect yourself from an approaching tsunami if you are in its path. (But with enough advance warning you could run uphill, or climb to the top of a tall, strong building.)

**4** The answer is: since it is exposed to the ocean, the UK could suffer a tsunami – and has done in the past. But we are not at high risk.

The UK is on the Eurasian plate, which is moving away from the North American plate, and pushing into the African plate. Usually, a tsunami is triggered by a major earthquake in the ocean floor. Major earthquakes generally occur only where plates push into each other, eg at the boundary with the African plate. (But a tsunami could also be triggered by an underwater landslide anywhere.)

The boundary with the African plate is currently not active with respect to earthquakes in the ocean, so the current risk is low.

If a tsunami approached the British Isles from the west, much of Great Britain's coast would be sheltered by Ireland. If it approached from the south, some coast would be sheltered by Ireland, and some by France.

**5** The buoys give out signals to satellites and these can transmit them very rapidly to places along vulnerable coastlines, giving people a chance to move away from the coast. (The buoys collect data from pressure sensors on the ocean floor, which can detect tiny changes in wave pressure.)

# What is a volcano?



## About this unit

This unit looks at the structure of a volcano, and describes the materials which may be thrown out during an eruption. It also describes the damage eruptions can do.

## Key points

- Volcanoes are where magma (melted rock) erupts through Earth's crust. Once out, the melted rock is called lava.
- The lava hardens to an igneous rock. (Basalt is by far the most common one.)
- Magma can be thin and runny, or thick and sticky (viscous).
- Thick sticky magma causes the most dangerous eruptions because it can't escape from a volcano easily; the pressure builds up until the magma explodes out.
- Explosive eruptions produce particles of all sizes, from dust to blocks of rock.
- A cloud of gas and particles from an explosive eruption will collapse, and rush down the slopes of the volcano as a deadly pyroclastic flow.

## Key vocabulary

volcano, erupt, magma, lava, viscous, vent, magma chamber, volcanic gas, pyroclastic flow, mudflow, crater, igneous rock, basalt, pumice

## Skills practised in 'Your turn'

- Geographical skills: q2, do a labelled drawing; q4, q5, interpret photos
- Literacy skills: q3, write an annotated list
- Thinking skills: q1, compare; q3, explain; q4, describe; q5, suggest reasons; q6, suggest strategies

## Unit outcomes

By the end of this unit most students should be able to:

- define the terms given in 'Key vocabulary' above
- draw a labelled cross-section showing the structure of a volcano
- explain why some eruptions are explosive, and others are 'runny'
- list the products from eruptions, and say what damage they cause

## Ideas for a starter

1. Shake a bottle of fizzy water vigorously. Ask: What will happen if I open the top? Why? (The gas will propel the water out of the bottle.) Open it and watch. The same thing happens in a volcanic eruption – but this time the gas is dissolved in magma. Some magma is very thick and gassy, so it explodes out.
2. With books closed, show a set of images of eruptions of all types, and from volcanoes of all shapes – classic conical volcanoes, fissures in the ground, and underwater volcanoes. Include eruptions seen from satellites. Ask for comments.

## Ideas for plenaries

1. If students have already studied Chapter 1 of *geog.3*, ask: When lava hardens, which type of rock does it form – igneous, metamorphic, or sedimentary? Link back to that chapter.
2. Search the Internet for *Which volcanoes are erupting today?* Display the map, and discuss.
3. If you did not choose starter 2, you could show the selection of images here.

## Further class and homework opportunities

Suggestions 14 and 15 on page 129 of this book

*geog.3 workbook*, page 47

**kerboodle**

### Resources

5.7 Sort it out! – Students label a diagram of a volcano.

5.7 You decide! – Students rank the hazards of a volcanic eruption.

5.7 Foundation worksheet – Students detail why pyroclastic flow is so deadly.

5.7 Extension worksheet – Students categorise the primary and secondary impacts of volcanoes, and explain why it is easier to predict a volcanic eruption than an earthquake.

### Assessment

5.7 End-of-lesson test – An assignable, auto-marked formative assessment.

## Answers to 'Your turn'

1 Magma is melted rock mixed with gas which is located underground, whereas lava is the name given to the same substance when it rises above ground level.

2 Note that in the drawing the cloud above the volcano is not yet a pyroclastic flow – it has not started to flow down the slope! Students could label it *cloud of ash, dust, bits of rock,* and *hot gas* or similar.

3 **a, b** There is no definitive 'correct' order since it depends on the nature and intensity of an individual eruption, and the area affected. A possible order is shown below. Generally, a pyroclastic flow will do more harm than lava flow.

increasing danger (↑)

- **pyroclastic flow**: very fast; scorches, burns and smothers everything in its path
- **ash**: can suffocate people and animals; destroys crops; causes buildings to collapse
- **volcanic gases**: cause acid rain; can kill people, animals and plants
- **lava flow**: destroys crops; can bury and burn towns and villages (but you can walk out of its way)

**c** Students could underline ash, because an ash cloud can spread quickly and is a major hazard to aircraft (after the 2010 Eyjafjallajökull eruption, flights were grounded for 6 days); or volcanic gases, as they may cause acid rain which could fall in other countries.

4 Students should describe the landscape in photo D. For example, the ground is covered in thick, white / grey ash; clouds of dust swirl in the wake of other rescuers. Some may also imagine the damage they can see to the small villages on the slopes of the volcano.

5 People may move to safety for a while in case of further eruptions. Once a volcano becomes active, there can be many volcanic events in succession. Since 2010, Mount Sinabung has erupted several times, so people would have been wise to stay away. Moreover, ash covering the farmland would have destroyed the crops and made the land economically unviable in the short term. (However in the long term, ash makes the ground more fertile.)

6 Two from: scientists can use a range of techniques to detect an eruption before onset (seismometers, thermal imaging, gas samples); the government could set up an early warning system for evacuation (radio, tv, text messages); educate people of the risks; run evacuation drills; set up an exclusion zone around the volcano; set aside an emergency fund; improve access for emergency services.

# 5.8 The most dangerous volcano in the world



## About this unit

In this unit students will find out why Mount Vesuvius is such a dangerous volcano.

## Key points

- Vesuvius is a highly dangerous volcano, because it is fed by viscous, gassy magma (which can lead to explosive eruptions and pyroclastic flows), and it is in a highly populated area.
- Vesuvius has erupted in the past and will erupt again, so it is monitored 24/7.
- The eruption in 79 CE destroyed Pompeii, killing its inhabitants, and covering it in 5 metres of ash and pumice.
- Excavations began in the 18th century to uncover Pompeii, and it is now a major tourist attraction.
- There are plans in place to evacuate people when there are any signs of a future eruption, but many have concerns about the plans.

## Key vocabulary

evacuate

## Skills practised in 'Your turn'

- Geographical skills: q1, q3, interpret maps; q4, use map scale
- Literacy skills: q3, q4, use a glossary; q6, evaluate; q6, justify
- Thinking skills: q2, give reasons; q3, q5, q6, explain; q6, make judgements based on various comments; q6, suggest strategies

## Unit outcomes

By the end of this unit most students should be able to:

- explain the term given in 'Key vocabulary' above
- explain why volcanoes are found in Italy
- give two reasons why Vesuvius is such a dangerous volcano
- explain how, and why, Vesuvius is monitored for signs of an eruption
- identify two actions that planners could take immediately to help a future evacuation of the area around Vesuvius to go more smoothly

## Ideas for a starter

1. With books closed, give students 5 minutes to find out as much as they can about Vesuvius. They could work in pairs. Create a spider map of their findings.
2. Show students the satellite image of Vesuvius and the surrounding area. Ask: What does this image show? What is Mount Vesuvius? Look at the built up area. Why is this a problem? What is Pompeii?

## Ideas for plenaries

1. Mount Vesuvius is about to erupt! Tell students they have to evacuate. Where will they go? How will they get there? What will they take? Ask them to write a blog, text, email, or draw an image to show how they feel.
2. Choose several students to take on the role of planners. Other students can ask them (sensible) questions about the evacuation plans for Vesuvius.

## Further class and homework opportunities

Suggestions 16 and 17 on page 129 of this book

*geog.3 workbook*, page 48

kerboodle

### Resources

5.8 Quiz time! – Students complete a quick quiz about Mount Vesuvius.

5.8 Word scramble – Students complete a crossword of key terms about volcanic eruptions.

5.8 Foundation worksheet – Students give examples of what can be done to predict, prepare for and protect against future eruptions of Mount Vesuvius.

5.8 Extension worksheet – Students write a structured answer evaluating whether the 'three Ps' of hazard management can be used to manage the threat of Mount Vesuvius' eruption.

### Assessment

5.8 End-of-lesson test – An assignable, auto-marked formative assessment.

## Answers to 'Your turn'

**1** Europe, Italy, Naples, Gulf of Naples / Mediterranean Sea

**2** Mount Vesuvius is considered an extremely dangerous volcano due to the nature of its magma – thick, sticky, and gassy, causing explosive eruptions. This could be deadly to the 600 000 people living within close range of its crater.

**3 a** A complex volcano is a landform made up of several related volcanoes. Campi Flegrei is a complex volcanic site composed of a network of craters and hydrothermal features.

**b** Italy has volcanoes because a destructive plate margin runs through it. The African plate is sinking under the Eurasian plate, leading to volcanoes (and earthquakes).

**4** About 8km. Students can use the scale on photo B to find this distance.

**5** Vesuvius is monitored closely because it has erupted at intervals over thousands of years and scientists know it will erupt again. An eruption could kill thousands of people if swift action is not taken to evacuate the affected areas.

**6 a** To evaluate the statement, students should judge how sensible it might be, considering both sides of the argument. Clearing homes off the slopes would reduce the hazard risk and could save lives. However, people who live there would protest as the volcano doesn't erupt regularly and they might prefer to take the risk. Some work in agriculture or tourism and profit from the volcano. Where would these people go and how would they earn a living if they were moved?

**b** Evacuating everyone by air would solve the issue of Naples' narrow winding streets, which would get blocked by traffic if people were told to leave the city en masse. However, Naples only has one airport, so students should think about its capacity and the potential for overcrowding. How would residents get there with all their belongings? How many planes could the government charter in an emergency?

**c** 'Move everyone by boat' (8) is the only other suggestion in D. Students might support this idea because Naples is by the sea and there is a large port shown in the photo. However, this could create chaos at the port, and it would lead to similar access issues as using only the airport. It would be more sensible to use a variety of transport modes to get people out of the city.

**d** Answers will vary.

**e** Answers will vary. Suggestions could include: widen roads where possible; improve traffic flow using one-way systems; improve the alert system; run evacuation drills; make city zoning clear so residents know whether they need to evacuate.

# So why live in a danger zone?



## About this unit

This unit looks at why people continue to live in areas which are prone to earthquakes, or with active volcanoes – despite knowing the risks.

## Key ideas

- It is estimated that just over 8% of the world's population live in areas of tectonic activity, where they are at risk from earthquakes and / or eruptions.
- There are many reasons why people continue to live in danger zones.
- Economic reasons: for example they have well-paid jobs, own land with fertile soil, cannot afford to move, are unlikely to find work anywhere else.
- Social reasons: for example they are familiar with the way of life there, enjoy life there, are optimistic that disasters will not occur or that they will be warned / protected, etc.

## Key vocabulary

quake-proof, tourism, geysers, fumaroles, geothermal energy, veins

## Skills practised in 'Your turn'

- Geographical skills: q1, q3, interpret and compare maps; q9, label a sketch
- Literacy skills: q4, summarise answer as a spider map; q5, q6, define
- Thinking skills: q2, q3, give reasons; q3, q6, explain; q4, decide then justify; q7, q8, come up with examples; q9, come up with a heating solution

## Unit outcomes

By the end of this unit most students should be able to:

- explain the terms given in 'Key vocabulary' above
- give two economic and two social reasons why people continue to live in danger zones
- give three examples of the benefits that volcanoes bring

## Ideas for a starter

1. With books closed, display a photo of a town or city below a lightly-erupting volcano, or people farming below a lightly-erupting volcano. Say: This is an active volcano. It could have a massive eruption at some point in the future. Why don't people move away? Write reasons on the board. You can return to them later.
2. Display a map of world population density and major cities (say over 5 million people). Compare with the plate map on page 91 of the students' book. Which cities / densely populated areas are at risk from tectonic activity (ie near plate edges)?

## Ideas for plenaries

1. Write *geothermal energy* on the board. Discuss what it means. Does it use natural resources? Is it renewable?
2. If you chose starter 1 you could return to it now. Students could modify or add to the reasons on the board.
3. If you did not do plenary 3 on page 116 of this book, or plenary 2 on page 122, search the Internet for *Where are earthquakes / volcanoes erupting today?* Display the map(s), and discuss locations / magnitudes.

## Further class and homework opportunities

Suggestions 18 and 19 on page 129 of this book

*geog.3 workbook*, page 49

**kerboodle**

### Resources

5.9 Where in the world? – Students identify the cities in a danger zone.

5.9 Word scramble – Students identify key terms about our restless planet.

5.9 Foundation worksheet – Students sort statements describing the aftermath of a volcano eruption, earthquake or tsunami to explain how this might differ in a rich or poor country.

5.9 Extension worksheet – Students create an advert selling a house located near to a volcano, including the benefits of buying in such a location.

### Assessment

5.9 End-of-lesson test – An assignable, auto-marked formative assessment.

## Answers to 'Your turn'

**1** These capital cities are close to plate edges so are at risk, mainly from earthquakes: Asmara (Eritrea), Guatemala City (Gautemala), Islamabad (Pakistan), Ankara (Turkey), Jakarta (Indonesia), Kabul (Afghanistan), Kampala (Uganda), Kathmandu (Nepal), Managua (Nicaragua), Manila (Philippines), Mexico City (Mexico), Panama City (Panama), Port-au-Prince (Haiti), Quito (Ecuador), San Jose (Costa Rica), San Salvador (El Salvador), Santiago (Chile), Tehran (Iran), Tokyo (Japan), Wellington (New Zealand).

**2** For example:

**a** you have a well-paid job; it's expensive to move; you might not find a new job elsewhere; you own fertile farmland in a volcanic area; you have cheap energy (in Iceland)

**b** hard to leave life and friends you know; hard to leave a comfortable lifestyle; people usually don't think disasters will happen to them; reassured by advances in monitoring, warning systems, construction, and by educational campaigns about how to respond in emergencies, and so on.

**3 a** Japan has so many earthquakes because it is located on a destructive plate boundary where several plates meet, including the Pacific plate and the Philippine plate.

**b** Rollers and flexible materials give buildings the ability to absorb shock and move with an earthquake's tremors. For example, flexible pads built into the foundation of a building isolate the base from the rest of the structure – so the base moves, but the section above ground remains steady.

**4** The big question is: where would everyone go? People usually get moved out only when danger is imminent, or after a disaster has already occurred. Even then the move may only be temporary, until damaged buildings are made safe, safer buildings constructed, etc.

**5 a** A fumarole is a vent or opening in Earth's crust which emits steam and gases

**b** A geyser is a spring of water which shoots into the air every so often; hot rock below ground heats it, and the pressure builds up

**6 a** Geothermal energy is generated from heat stored in rock. Water is pumped down into the hot rocks, and it comes back up as steam. This is used to heat homes or drive turbines to make electricity.

**b** This power station does not contribute to climate change because unlike burning fossil fuels, generating geothermal energy does not have a huge carbon footprint. $CO_2$ is not released during this process.

**c** Iceland is situated on the Mid-Atlantic Ridge, which is a constructive plate margin, so plates are moving apart. Melted rock (magma) rises to plug the gap, bringing heat from Earth's mantle.

**7** Students might mention things like:

- fossil fuels used for electricity generation, home heating
- oil used for transport and as a source of chemicals for plastics, pharmaceuticals, etc.
- jewellery made of gold and silver

Note that most mined metals are associated with magma formed at plate boundaries, where one plate has dived down.

**8** Some examples: people in demolition and construction industries (demolishing unsafe structures, building new ones); developers (buying up cleared land); suppliers of timber, metal, and other building materials; architects and engineers designing earthquake-proof buildings; scientists studying earthquakes (they can learn more from each one).

**9** Students could show river water being pumped down to the hot rock, and then led to the homes (as hot water or steam).

# Our restless planet



## Skills practised in check questions

- Geographical skills: q1, q2, interpret a diagram; q3, interpret a map; q3, compare maps; q4, interpret a photo
- Literacy skills: q6, extended writing
- Numeracy skills: q4, interpret a scale
- Thinking skills: q1, q3, q4, name geographical features; q1, q2, q3, q4, explain; q1, compare; q2, q4, q5, give reasons; q4, suggest solutions; q6, comment on the extent of an issue

## Answers for check questions

**1 a i** Mantle

**ii** Core

**b** The crust

**c i** R is the lithosphere. Notice that it contains both the crust and the thin layer of the upper mantle

**ii** Plates

**d** The rock at S flows – but very slowly

**e i** Continental crust

**ii** Oceanic crust

**f** U (oceanic crust) is heavier

**2 a** Answer is **i** – they are converging

**b** At the point where these plates meet, the African plate is oceanic, meaning it is denser and heavier than the Eurasian plate, which is continental at this location under Italy.

**c** The magma comes from the African plate because it is melted by Earth's mantle as it sinks.

**d i** Scientists want to find out when the next eruption of Vesuvius is likely to be. This is because the city of Naples (and its surrounding urban areas) are extremely close to this very dangerous volcano. Many lives would be at risk if the people were not given enough time to evacuate before an eruption.

**ii** The temperature inside a volcano increases before an eruption because hot magma begins to rise to the surface.

**e** Italy is located on a destructive plate boundary

**f** The UK was located on a plate margin in the past but has not been close to one for millions of years, hence no active volcanoes.

**3 a** The Mid-Atlantic Ridge

**b** The North American and Eurasian plates meet in the northern hemisphere

**c i** Volcanic eruptions occur because as the plates are diverging, magma rises between them from the lower mantle.

**ii** Earthquakes occur because the movement of the plates over the mantle creates friction.

**d** Basalt

**4 a** Earthquakes are caused by any sudden large movement of rock. This has the greatest effect at plate boundaries, but the shockwaves can travel long distances, so countries like the UK sometimes feel them. Tectonic activity is not the only cause of earthquakes; the collapse of a mine shaft can cause one, and fracking has also been linked to small earthquakes.

**b** Because pressure often builds up at plate margins, due to the movement of plates over the mantle.

**c** Seismic waves

**d i** Richter scale

**ii** 9.2

**e** Buildings are the main reason for earthquake deaths; when they collapse, people are crushed inside. Buildings in developing countries are less likely to withstand the tremors due to lower emphasis on building codes and the higher cost of aseismic buildings. Other reasons: population density; lack of preparedness; poor governance; or the fact that large urban areas are often located on coasts, which correspond with many of Earth's plate boundaries.

**f** Building codes can be used so that new buildings are designed to withstand earthquakes. In Chile, all new buildings are designed to withstand a 9.0 earthquake. However, it is often too expensive to rebuild dangerous older buildings. Space between buildings for people to gather safely would also save lives. People can be educated on what to do in the event of an earthquake; this includes earthquake drills.

**5** Two reasons from: most settlements formed before scientists understood tectonic hazards and it would be too difficult to move them now; people have formed attachments to these places; people have jobs in these locations; earthquake-proof building designs are getting better so there is less risk. If volcanoes are also present, there are other benefits such as: good soil for farming; money from tourism; geothermal energy; valuable materials found in extinct volcanoes.

**6** For this extension question, students are asked to state the extent to which they agree with a statement. Good answers will give evidence to support both sides of the argument and provide a short conclusion.

An example answer: There are many ways to protect people from earthquakes and volcanoes. Scientists can now predict volcanic eruptions in advance and people can be evacuated, which saves lives. Earthquakes are harder to predict, but building designs have improved and in earthquake zones, many buildings are designed to withstand large earthquakes.

However, although it is now easier to prevent deaths, it is still difficult to protect places from damage. Earth's population is increasing and urban areas are growing. Giant tsunami waves can wreck miles of coastline in just a few minutes. Poorer countries may not be able to afford early warning systems and they may suffer outbreaks of disease in the aftermath of a disaster. Overall, the extent to which we can limit the harm caused by a natural hazards depends on factors such as the type of hazard and the level of development of a country.

# Further suggestions for class and homework

Most of these suggestions are addressed to your students. Where research or further resources are needed, the Internet will almost certainly provide the answer.

The suggestions are graded *, **, *** according to level of difficulty. Some are suitable for all levels, and can be differentiated by outcome.

## What's Earth like inside?

1 **Your Earth** Design a colourful image of the inside of Earth, to go on a T-shirt or birthday card. Add a slogan / message! */**/***

2 **Capture that heat!** Design a system for using the heat inside Earth to heat homes and / or cook food, all over the world. **/***

3 **Ocean mountains and trenches** If you drained the oceans you'd find amazing mountains and trenches. Name the tallest mountain in the ocean, and the deepest trench. Mark them on an outline world map. Then write an exciting paragraph about each. Add diagrams and photos. **/***

## A cracked Earth?

4 **On the edge** Use a world political map and a plate map to name 15 countries that lie on, or close to, plate edges. */**/***

5 **The fossil mystery** The rocks in Britain contain many fossils from warm seas around the Equator. So how did they get here, over 50° north of the Equator? Prepare a presentation for a group of 11-year-olds, to explain this mystery. Unit 1.5 may help. */**/***

## A closer look at plate movements

6 **Ocean trenches** What are they like? What kinds of creatures live in them? Find out and tell the class. You could take the Mariana Trench, the deepest one, as your example. You may find an animation of a fly-by, on the Internet. **/***

7 **Demonstration: fold mountains** Fold mountains form when plates push into each other, folding rock upwards. Think of a way to demonstrate this. (Use dough? paper napkins? cloth?) Make sure it works, and show the class. */**

## What is an earthquake?

8 **The San Andreas fault** Find out: where it is; which plates meet at this fault; about one earthquake caused by the fault. Write an article for a geography magazine. **/***

9 **Which country had an earthquake today?** (If not used in class already.) Try an Internet search for *earthquakes today*. See if you can download a map, or else mark the earthquake sites on an outline map of the world. ***

## An earthquake in Southwest China

10 **Help!** It is 2008 in Sichuan. Your job is to organise people to help, after the earthquake. What help will be needed? Doctors? Search teams? Sniffer dogs? Tents? Those are just the start! Write a full list. Then write a formal plea to the governments of other countries, requesting help. */**/***

11 **Earthquake prone** It is 2008 in Sichuan, after the earthquake. You are an official. Write a formal report for the government explaining why Sichuan will continue to have earthquakes, and why the building regulations must be tightened up. You could include a plate map. **/***

## Tsunami!

12 **Launch an appeal for disaster victims** It is 28 December 2004. You have a 40-second TV slot to launch an appeal for help for the survivors of the tsunami. Write a script to fit exactly 40 seconds. Draw a storyboard for it, showing the images you will broadcast. */**/***

13 **A tsunami warning system for the Indian Ocean** There is a warning system in place now, as a result of the 2004 tsunami. So here's a challenge: see if you find out how it works, and then explain to the class, in simple words! ***

## What is a volcano?

14 **Make a volcano** You will need strips of newspaper, glue, and a small plastic bottle as a base. Dip the strips in the glue and wrap them around the bottle to build up a volcano shape. Add one or two secondary vents. When dry, paint your volcano with water-based paint, and coat with varnish. *

15 **Adopt a volcano** Choose an active volcano anywhere in the world. For example, Popocatépetl in Mexico, or the Soufrière Hills volcano on Montserrat in the Caribbean. (An Internet search for *active volcanoes today* will help.) Mark its location on a world map, explain why it's there, and keep a diary of its activity. Check twice a week? */**

## The most dangerous volcano in the world?

16 **Bring in the tourists** You have to design and write a fold-out brochure enticing tourists to visit Pompeii. So first, find out more about it. Write a list of points to include in your brochure. Then create it! (You could work in pairs.) */**/***

17 **Puzzle them!** Create a crossword puzzle with clues, or a set of jumbled words, or a set of sentences with missing words and letters, about what you have learned about volcanoes. Give it to a partner to try! See how difficult you can make it. */**/***

## So why live in a danger zone?

18 **Quake-resistant building** How are buildings made quake-resistant? Try an Internet search. Draw sketches or obtain photos of buildings, and label the special features. ***

19 **Predicting earthquakes** If we could predict earthquakes, hundreds of thousands of lives would be saved. But how good are experts at predicting them? Work with a partner or small group to find out. Start by typing *predicting earthquakes* into a search engine? ***

# 6 About Russia

## About this chapter

### Addressing the KS3 Programme of Study

- Extend locational knowledge and deepen spatial awareness of the world's countries, using maps of the world to focus on … Russia, its environmental regions, key physical and human characteristics, countries, and major cities

### About the content

- This chapter starts with an overview of Russia and its history.
- It looks at Russia's physical and human geography, and takes a tour of European Russia.
- It ends by looking at climate change in Yakutia, Siberia, and how Russia is responding to the possibility of an ice-free Arctic Ocean.

## The big picture

These are the key ideas in this chapter.

- Russia is the world's biggest country by area. It lies partly in Europe, but mostly in Asia. About 78% of its population live in Europe. (The Ural Mountains mark the divide.)
- Russia was once the major part of the vast Russian Empire, with its beginnings around 1300. After a revolution in 1917, the empire became the Union of Soviet Socialist Republics (the USSR, or Soviet Union). This broke up in 1991.
- Much of Russia has a severe climate, under snow for up to eight months of the year. As a consequence, much of it is empty of people. The climate is milder in the European part.
- Climate change is having a major impact, as Russia is warming two and a half times faster than the global average.
- Russia is preparing to exploit the Arctic Ocean as more ice melts each summer.

## The chapter outline

Use this to give students a mental roadmap for the chapter.

**6** **About Russia** The chapter-opening photos in the students' book are an important resource. See opposite for using them.

**6.1** **Meet Russia!** An overview of Russia, and a little history

**6.2** **What are Russia's main physical features?** Explores the main physical features based on a map

**6.3** **Russia's climate zones and biomes** Get a feel for the cold – and meet the vast taiga and Russia's other biomes

**6.4** **What about the Russian people?** How the population is concentrated in European Russia, where the climate is milder

**6.5** **A tour of European Russia** An overview of European Russia

**6.6** **Climate change in Yakutia** How rapid warming is affecting Yakutia

**6.7** **What is Russia doing in the Arctic Ocean?** How Russia is responding to the likelihood of an ice-free (in summer) Arctic Ocean

**6.8** **About Russia check questions** A page of questions to check knowledge, understanding, and skills across the chapter, with an opportunity for extended writing

## Objectives and outcomes for this chapter

| Objectives | Unit | Outcomes |
|---|---|---|
| Most students will understand: | | Most students will be able to: |
| • what and where Russia is | 6.1 | • show Russia on an outline map of the world; say it's the world's largest country by area, and partly in Europe but mostly in Asia, with the Urals as the dividing line; name seven countries which border Russia; name the oceans it borders; outline Russia's history from 1300 CE |
| • that Russia has a wide range of physical features | 6.2, 6.5 | • name and locate on a map the physical features listed in Unit 6.2, plus the Kola peninsula |
| • that Russia has a range of climates – and corresponding biomes | 6.3 | • describe, Russia's climate zones and biomes and explain the link between them; explain why winters are so cold in most of Russia; explain what permafrost is and where it's found |
| • that Russia's population distribution is influenced by climate | 6.4, 6.5 | • describe and explain the pattern of population density; name and locate on a map the two biggest cities (Moscow and St Petersburg) plus Sochi, Murmansk, Kaliningrad, Yakutsk, Vladivostok; say that there is great inequality in Russia |
| • how climate change is affecting Yakutia in Siberia | 6.6 | • explain why Yakutia is of great importance to Russia; how climate change is affecting Yakutia |
| • the opportunities that Russia sees as summer ice in the Arctic decreases | 6.7 | • explain how the melting ice can benefit Russia; how Russia is getting ready to exploit the Arctic Ocean |

These tie in with the outcomes for each unit in this teacher's handbook.

## Assessment for this chapter

There are assessment materials for this chapter on *geog.3 Kerboodle Lessons, Resources, & Assessment*. They include interactive auto-marked end-of-lesson and chapter-level tests, a printable exam-style end-of-chapter paper (with a mark scheme), and a printable self-assessment checklist. See also the notes on pages 14–17 of this book.

## Getting ready for this chapter

There is a scheme of work for this chapter, plus a glossary and glossary worksheet, on *geog.3 Kerboodle Lessons, Resources, & Assessment*.

## Using the chapter-opening photos

Ask students:

These photos are all about Russia. Which continent is Russia in? Are you sure?

What can you say about Russia from these images?

What else do you know about Russia?

The photos on page 107 of the students' book show in order, row by row:

- St Basil's Cathedral in Moscow
- a bandy match; a form of ice hockey, bandy is a very popular sport in Russia
- selling fish at a market stall in Yakutsk, Siberia (–30 C in winter)
- a satellite image of Russia, and the Russian flag
- a gas pipeline heading for a power station
- a typical small village in central Russia
- a car worker at a car assembly plant near St Petersburg.

# Meet Russia!



## About this unit

This unit introduces Russia – with maps, and some key facts.

## Key points

- Russia is the biggest country in the world, by area, and so wide that it has 11 time zones.
- It lies on two continents: Asia and Europe, with the Asian part east of the Ural Mountains. Most of its land is in Asia – but most of the population is in Europe.
- Russia was once a vast empire, the Russian Empire. This collapsed in a revolution in 1917, and the territory was reorganised to form the Union of Soviet Socialist Republics (the USSR, or Soviet Union), controlled from Moscow.
- The USSR broke up in 1991, with the republics becoming independent countries bordering Russia.
- Crimea is part of Ukraine, but Russia controls it.
- Russia is rich in natural resources and earns money by exporting these.

## Key vocabulary

Bering Strait, exclave, USSR, Soviet Union, Muscovy, Russian Empire, Russian Revolution, republic, communism

## Skills practised in 'Your turn'

- Geographical skills: q1, q2, q3, q5, q6, q7, interpret a map; q8, interpret a pie chart
- Literacy skills: q4, define; q8, use a glossary
- Thinking skills: q4, explain; q8, use evidence to support a viewpoint

## Unit outcomes

By the end of this unit most students should be able to:

- explain the terms given in 'Key vocabulary' above
- locate Russia on an outline world map, and indicate the Arctic Circle running through it
- say that Russia lies partly in Europe, but mostly in Asia
- name at least seven countries which border Russia, including China

## Ideas for a starter

1. With books closed, give out atlases. Play a 'beat the teacher' game against the clock. You ask questions about Russia, and students must answer within a given time. If they fail, or get the answer wrong, 1 point to you. Questions could include: Name a country beginning with K which borders Russia. Which continent is Russia in?
2. Show a set of very varied images of Russia. Ask what impression each image gives. Would a different set of images leave you with a different overall impression?

## Ideas for plenaries

1. Look at map A. Why do we show it that way round?
2. What is a time zone?
3. Why study Russia?

## Further class and homework opportunities

Suggestions 1–4 on page 147 of this book

*geog.3 workbook*, page 50

**kerboodle**

### Resources

6.1 Quiz time! – Students complete a quick quiz about Russia.

6.1 Where in the world? – Students complete a quick quiz about Russia's borders.

6.1 Foundation worksheet – Students cross out incorrect words in a paragraph on Russia, list former USSR countries, and explain how several keywords factor into Russia's history.

6.1 Extension worksheet – Students create a factfile on Russia, including location, capital city, physical and human geography, and key points in history.

### Assessment

6.1 End-of-lesson test – An assignable, auto-marked formative assessment.

## Answers to 'Your turn'

**1** **a** Asia and Europe

**b** Asia has the bigger share (by land area, but not by population)

**2** **a** **i** any three of: Kazakhstan, Mongolia, China, North Korea

**ii** any four of: Azerbaijan, Georgia, Ukraine, Belarus, Latvia, Estonia, Finland, Norway, plus Poland and Lithuania which border the exclave Kaliningrad

Note that there is dispute about whether Azerbaijan and Georgia are in Europe or Asia. They (along with Armenia) consider themselves European. They belong to the Council of Europe.

**b** Kazakhstan

**3** Arctic Ocean and Pacific Ocean

**4** **a** Part of a country that's cut off from the main part; you have to cross the territory of another country (or countries) to reach it

**b** Most countries say Crimea is not part of Russia because under international law it is still part of Ukraine. In 2014, residents voted for it to become part of Russia, but this was condemned as an illegal move by The United Nations, which is a global organisation made up of all but two of the world's countries (Russia and China).

**5** **a** 82 km

**b** The Bering Strait

**6** The answer is **c**, about one fifth of Russia lies within the Arctic Circle.

**7** **a** In 1300, the state of Muscovy was very small because Moscow's control was in its infancy. Over centuries the area it controlled grew and expanded out from Moscow across Asia and Europe in all directions, taking up land which is now part of other countries. By 1721, Russia became so large that it was renamed 'the Russian Empire.'

**b** In 1990, the Russian lands were being ruled by the Communist party from Moscow. This territory was called the USSR. The states within it voted to break up in 1991 which meant that many were no longer controlled by Moscow. Moscow's reduced territory became known as the Russian Federation and was smaller than the USSR had been. By 2020, it had not changed much.

**8** True – most of Russia's exports are primary products. C shows that in 2020 over three quarters of all exports were primary. These include: oil and oil products, gas, coal, metals, timber, and wheat.

# 6.2 What are Russia's main physical features?

Help at a glance

## About this unit

This unit provides details about Russia's physical geography.

## Key points

- Much of Russia is low and flat. The East European and West Siberian plains, separated by the Ural Mountains, cover vast areas. The huge Central Siberian Plateau is more elevated.
- The Ural Mountains form a natural border between Asia and Europe. The Caucasus Mountains include Mount Elbrus, Europe's highest mountain.
- In Siberia several very long rivers flow from south to north, including the Ob, the Yenisey and the Lena. The Amur forms a natural border with China, while the Volga is Europe's longest river.
- Russia's Kamchatka Peninsula and Kuril Islands have many active volcanoes.

## Key vocabulary

plain, plateau, peninsula, Ring of Fire

## Skills practised in 'Your turn'

- Geographical skills: q1, q2, q4, q5, interpret a map
- Thinking skills: q2, describe; q3, list facts; q5, explain; q6, create a spider map to summarise knowledge

## Unit outcomes

By the end of this unit most students should be able to:

- explain the terms given in 'Key vocabulary' above
- name and locate the physical features mentioned in this unit, on a map of Russia

## Ideas for a starter

1. With books closed, display a satellite image of Russia. How much can the class deduce from it about Russia's physical features?
2. With books closed, ask: What are *physical features*? Then give out atlases. Students have five minutes to note down at least ten facts about Russia's physical features. They could work in pairs or small groups. Could the facts be grouped in some way? Are some more important than others?

## Ideas for plenaries

1. The Ural Mountains make a good boundary. Why? (straight, not too wide)
2. What's the difference between a plain and a plateau?
   (Plain: low flat area. Plateau: elevated fairly level area.)
3. So the world's deepest lake and largest lake are both in Russia. Which is which?
4. What is the Ring of Fire? (before turning to page 90!)
5. Is the Arctic Circle a physical feature?
6. It is easy to mix up the Caspian, Baltic, Black, and Bering Seas, and Lake Baikal. See if you can work out a good way to remember which is which. The class can choose the best.

## Further class and homework opportunities

Suggestions 5–8 on page 147 of this book

*geog.3 workbook*, page 51

kerboodle

### Resources

6.2 Where in the world? – Students complete the descriptions of Russia's main physical features.

6.2 Word scramble – Students complete a crossword of Russia's physical features.

6.2 Foundation worksheet – Students create questions, fill out missing vowels, and complete a quiz on Russia's main physical features.

6.2 Extension worksheet – Students write questions about and describe Russia's physical features, and explain why someone would want to live on either a plain or plateau.

### Assessment

6.2 End-of-lesson test – An assignable, auto-marked formative assessment.

## Answers to 'Your turn'

1 **a** The Ural Mountains

**b** The Caucasus Mountains

**c** The West Siberian Plain

**d** The Kamchatka Peninsula

**e** Lake Baikal

2 We use the term *Siberia* to mean Asian Russia: all of Russia to the east of the Urals. This is the historical, and widest, definition. But note that there are other definitions. For example in 2002 the Siberian Federal District was established, covering a much smaller area. Another definition excludes some of southern Russia, and the extreme east.

3 Three from: it's salty; it's a huge lake; it's the world's largest lake in terms of area; over 130 rivers flow into it; the river Volga flows into it; no rivers flow out of it; it does not overflow due to evaporation; it makes up part of Russia's south-west border. If students look at the world map on page 140–141, they will see that it also forms part of the border of Iran, Kazakhstan, Turkmenistan, and Azerbaijan.

4 **a** The River Amur

**b** **i** Any five of the others that are labelled

**ii** The River Volga, or River Don

5 The River Lena often floods not due to excessive rain, but due to water bursting through ice dams which have formed in the river's lower course. These occur in early summer when ice has begun to thaw in central and southern Russia. It might be useful for the class to discuss this, looking at the map on page 110 for reference. The River Lena flows from south to north, with its mouth in the Arctic Ocean. This means that its upper basin is in a warmer climate than the lower part of its basin, which is in the Arctic Circle. In cold temperatures, whole rivers can freeze, and during summer when permafrost melts in Siberia, river levels rise.

6 Encourage students to use a large sheet of paper (A4 turned sideways, or two sheets of A4 taped to make A3), and then to continue to add to the spider map during the rest of the topic.

# Russia's climate zones and biomes



## About this unit

This unit explores the link between climate and biomes, for Russia.

## Key points

- Climate everywhere is influenced by latitude, altitude, and distance from the sea.
- Much of Russia lies further north than the UK. About a fifth of it lies within the Arctic Circle. It is also a vast land mass, most of it far from the moderating effect of the sea.
- So, overall, Russia is a cold country and the tundra and sub-Arctic climate zones have permafrost – soil below the ground which is permanently frozen.
- Overall, precipitation is low. There's little evaporation so the air is dry, and mountains block moist air.
- A biome is a region with similar natural vegetation, and animals, across it. Biomes are closely linked to climate. Russia's biomes include tundra, taiga, temperate forest, steppe, and mountain.

## Key vocabulary

latitude, altitude, distance from the sea, tundra climate, sub-arctic climate, humid continental climate, semi-arid climate, mountain climate, permafrost, biome, taiga, steppe, temperate forest, tree line

## Skills practised in 'Your turn'

- Geographical skills: q2, interpret maps; q5, compare maps and identify patterns
- Literacy skills: q1, use a glossary
- Thinking skills: q3, explain, give reasons; q4, compare; q5, give evidence; q6, choose, and justify choice

## Unit outcomes

By the end of this unit most students should be able to:

- explain the terms given in 'Key vocabulary' above
- explain how latitude, altitude, and distance from the sea affect climate
- explain why Russia is so cold overall, and quite dry
- explain the link between climate and biomes
- compare two different biomes

## Ideas for a starter

1. What key line of latitude crosses Russia? So what can you guess about Russia's climates?
2. Show a set of images or video snippets related to climate in Russia, from warm / hot and sunny in the south-west to bitter cold in Siberia. What conclusions can the class draw?
3. Find on the Internet and display maps of average air temperature in January and July, for Russia. In particular, the January map is very dramatic. Imagine −48 °C!

## Ideas for plenaries

1. You could use questions 1, 2, and 3 of 'Your turn' as plenaries.
2. Then ask further questions about map A. For example: What's the climate like in Moscow? Around Lake Baikal? Just north of the Caspian Sea?
3. If you did not use starter 3, you could use it as a plenary.
4. I packed my bag for a trip to Lake Baikal in winter, and I took ...

## Further class and homework opportunities

Suggestions 9–11 on page 147 of this book

*geog.3 workbook*, page 52

**kerboodle**

### Resources

6.3 Sort it out! – Students identify the features of Russia's biomes.

6.3 Where in the world? – Students locate Russia's biomes on a map.

6.3 Foundation worksheet – Students complete a table with words and pictures on the five biomes in Russia, outlining the animals and trees / plants that appear in each region.

6.3 Extension worksheet – Students fill out a table detailing features of Russia's biomes, and justify selecting two biomes to appear in a TV programme about beautiful parts of Russia.

### Assessment

6.3 End-of-lesson test – An assignable, auto-marked formative assessment.

## Answers to 'Your turn'

**1** **a** Generally, the further north or south of the Equator you go, the cooler it gets.

**b** The higher you go above sea level, the cooler it gets.

**c** The sea warms more slowly than land does, in summer, and cools more slowly in winter. This in turn affects the climate near the coast. Inland, away from the sea, land heats up fast in summer, and gets hotter, than land at the coast. Likewise it cools faster in winter, and gets colder, than land at the coast.

**2** **a** Sub-arctic **b** European Russia

**3** **a** Much of the precipitation will be in the form of snow.

**b** There won't be that much evaporation from the cold Arctic Ocean. Moist winds from other seas would lose moisture on the long journey to inland Russia. Mountain ranges would block moist winds from elsewhere.

**4** Comparisons could include: the taiga is mostly coniferous forest, whereas the steppe is mostly grassland; permafrost is common in the taiga biome, but not in the steppe; the taiga is a huge biome, but the steppe covers a much smaller area; the land is better for farming in the steppe biome than the taiga; different wild animals live in the two biomes; the taiga spans from east to west, but the steppe is only found in the south-west of Russia.

**5** Climate is the key factor, for biomes. Students should notice that patterns on the map match quite well in places. For example:

- the sub-arctic climate zone is covered in taiga
- the tundra climate zone matches the tundra biome closely
- the semi-arid climate zone has a steppe biome
- an area with a mountain climate in south-west Russia, on map A, has a corresponding mountain forest / mountain steppe biome

The match is least obvious for the humid continental zone, in which the temperate forest gives way to taiga, as you go east.

**6** We need food, water, and a certain level of warmth. Students should conclude that the tundra would be very cold to live in, and it would be impossible to grow crops – so you'd be limited to hunting and fishing for food, and melting ice for water. The climate is also very harsh in winter in most of the taiga, and there's permafrost, and summers are short, making farming very difficult. In the mountain biome it's cold, and vegetation is sparse, so there's not much scope for farming there either. That leaves the temperate forest and steppe biomes to choose from. Students could look at the population density map in Unit 6.4 to see where humans have ended up living.

# 6.4 What about the Russian people?

Help at a glance

## About this unit

This unit looks at Russia's population – where people live, and how well off they are.

## Key points

- Russia has a population of around 146 million, or about 2.3 times larger than the UK.
- Population is unevenly distributed.
- Around 78% of the population live in European Russia, where the climate is milder.
- Thanks to its history, Russia has over 185 different ethnic groups. The Russian ethnic group is the biggest (81%).
- The population is mainly urban (75%), and 67% of the workforce is in the tertiary sector.
- There is great inequality in Russia. The richest 10% hold nearly 90% of the wealth, while 14% have hardly enough to survive (2019).

## Key vocabulary

ethnic groups, urban, fertility rate, inequality, oligarchs

## Skills practised in 'Your turn'

- Geographical skills: q1, q4, interpret a map; q2, compare maps; q2, q3, identify patterns on a map; q6, interpret pie charts
- Literacy skills: q5, define; q6, use a glossary
- Numeracy skills: q7, interpret and manipulate data from a table
- Thinking skills: q1, q3–q7, explain; q2, give evidence to support an answer; q3, q4, describe; q6, list; q7, compare

## Unit outcomes

By the end of this unit most students should be able to:

- explain the terms given in 'Key vocabulary' above
- describe the pattern of population distribution in Russia, and explain it
- name and locate on a map of Russia at least the five biggest cities (listed in the panel), and Vladivostok (at the end of the Trans-Siberian railway).
- describe the structure of employment in Russia, and say how it compares with the UK's
- compare GNI per person (PPP) for Russia, China, and the UK

## Ideas for a starter

1 With books closed, display a map of average air temperatures in January, for Russia. Imagine those temperatures. Might some places be too cold to live in? Might temperatures affect population density? Leave the map on display to allow students to compare it to the map on page 114 of their books.

2 Which well-known or famous Russians, living or dead, can you name?

## Ideas for plenaries

1 Why didn't they run the railway straight across the middle of Russia?

2 Of Russia's top five cities, how many are in European Russia?

3 So ... what would you say is the main factor influencing Russia's population distribution? To confirm that it is climate, compare the map of climate zones, or January air temperatures, with the population distribution map.

4 Search in Google Images for *Russia at night* to find a satellite image showing the lights of settlements.

## Further class and homework opportunities

Suggestions 12–14 on page 147 of this book

*geog.3 workbook*, page 53

**kerboodle**

### Resources

6.4 Sort it out! – Students choose the best location for a furniture factory.

6.4 Quiz time! – Students complete a quick quiz about Russia's people.

6.4 Foundation worksheet – Students analyse the urban population of Russia and compare countries' gross national income (GNI) per person.

6.4 Extension worksheet – Students round GNI-per-person values to the nearest thousand and draw and label a bar chart to display this information.

### Assessment

6.4 End-of-lesson test – An assignable, auto-marked formative assessment.

## Answers to 'Your turn'

**1 a** Most live in European Russia – mainly its southern half

**b i** Less than 1 person per sq km

**ii** 10–100 people per sq km

**2** Climate has a major influence over population density in Russia. The mildest regions are the temperate forest and the steppe, which are in the south-west. This correlates with the areas of highest population density on map A. The lowest population densities are in central, eastern and northern parts of Russia, which is mostly made up of the taiga and tundra biomes. These are very cold and are covered in ice for most of the year. On the other hand, there are some areas of higher population density on the map, which do not relate to climate. These are along the railway, along the rivers and near the coast.

**3 a, b** In Asian Russia, in the areas with fewer people, the density is generally greater along the rivers. People settle along rivers because they can use them for water supply, for food (fish), and for transport. Even when frozen, rivers are excellent for transport.

**4 a** Moscow in the west and Vladivostok in the east

**b** Again there appears to be quite a close relationship between the railway route and population density, especially in the eastern half. Settlements along the route, before the railway was built, would have influenced the choice of route. But it is very likely that the railway in turn helped to increase the population around stops along the route, because it gave people access to other places, for goods and services, and to sell things. It would have made life a lot easier.

**5 a** The replacement fertility rate is the average number of children per woman that keeps a population at the same size (without migration): it is taken as 2.1 children.

**b** Russia's population is likely to fall due to a low fertility rate of 1.6 babies per woman. At a fertility rate of 2.1, the population would remain stable. However, a country's population can also change due to migration. If more immigrants came to Russia, the population could rise.

**6 a** Russia has more workers in farming

**b** Service sector jobs: nursing, teaching, hospitality, accounting, driving, engineering, etc.

**c** Students' opinions will vary. Some may decide it is good to have industry because selling processed goods generates profit (countries such as China have become richer as a result of their large industrial sector). Others may suggest that it would be better to have more jobs in the service sector, as these provide higher average incomes than jobs in factories.

**7 a** GNI per person (PPP) – the GNI per person is adjusted to take into account that things cost more in some places than others.

**b** USA

**c** The UK's average income is the highest of the three, at \$49 040. This is more than double China's which is \$16 760. Russia sits in the middle at \$28 270. These figures could be explained by looking at the pie charts in B. Typically, countries with a small agricultural sector and a large services sector have a higher GNI. Students could find an equivalent pie chart online for China. (Notice that these figures are in dollars. It may be useful to look at current exchange rates and ask students to convert the data to GBP. The class could also discuss why the dollar is considered the dominant world currency).

**d** The average income for Russia is \$28 270, but this does not mean that all people earn the same. In fact, the richest 10% have 90% of the wealth which points to extreme inequality in Russia. At the other end of the scale, some Russians are living below the poverty line and struggle to survive. (To investigate this further, the class could look at figures online for the Gini Index. This is a measure of how fairly income is distributed in every country – a low score means it is more equal).

# A tour of European Russia



## About this unit

This unit takes a quick tour of European Russia west of the Urals.

## Key points

- European Russia – west of the Urals – comprises only about a quarter of Russia's land area, but has about 78% of its people.
- Here are some facts about European Russia:
  - Moscow is Russia's largest city, and St Petersburg is second largest.
  - Sochi, on the Black Sea, is Russia's top holiday resort.
  - European Russia has important strategic ports: Kaliningrad and St Petersburg on the Baltic Sea, and Murmansk on the Arctic Coast.
  - Crimea belongs to Ukraine, but is controlled by Russia.
  - The North Caucasus includes the republic of Chechnya.
  - There is a lot of farmland, but many rural villages are in decline.

## Key vocabulary

North Atlantic Drift

## Skills practised in 'Your turn'

- Geography skills: q1, interpret photos; q2, q6, describe locations; q3, identify geographical factors
- Thinking skills: q1, choose, and explain choice; q4, q6, explain; q5, suggest reasons; q7, identify benefits

## Unit outcomes

By the end of this unit most students should be able to:

- explain the term given in 'Key vocabulary' above
- name, describe, and locate on a map of Russia at least these places named in this unit: Moscow, St Petersburg, Sochi, Kaliningrad, Murmansk
- list the ports named in this unit and explain each one's importance to Russia

## Ideas for a starter

1. Russia gets millions of tourists every year. Which do you think most visit, European or Asian Russia? Why? Come up with as many reasons as you can.
2. Does Russia have any good football or other sports teams? Where are they based? Might there be any special problems in running a Russian football league? (Teams may have to travel huge distances, crossing several time zones; bitterly cold winters in places.)

## Ideas for plenaries

1. How many of the named places have you heard of? In what context?
2. Students choose one of the named places on the map to explore on Google Earth.
3. Show video footage for St Petersburg or Moscow.
4. Give out an outline map of European Russia. With books closed, students mark in and label the places and features you call out. Then they can swap with partners to review / score. Agree on a scoring system.

## Further class and homework opportunities

Suggestions 15–17 on page 147 of this book

*geog.3 workbook*, page 54

**kerboodle**

### Resources

6.5 Word scramble – Students complete a crossword of places in European Russia.

6.5 Sort it out! – Students sort the characteristics of places in Russia.

6.5 Foundation worksheet – Students use pages 116 and 117 of *geog.3* to answer questions on European Russia, matching places to descriptions and choosing two locations to visit.

6.5 Extension worksheet – Students answer questions on European Russia and produce a short extract for a travel magazine to attract UK visitors to Sochi.

### Assessment

6.5 End-of-lesson test – An assignable, auto-marked formative assessment.

## Answers to 'Your turn'

1 Answers will vary.

2 a It is in the North Caucasus, which lies along the northern side of the Caucasus Mountains, in the most southerly part of European Russia.

b It's the peninsula shaped like a fat sausage, in the north of European Russia. It juts into the Arctic Ocean (or more specifically, the part of the ocean called the Barents Sea), and extends from Russia's border with Norway down to the Arctic Circle.

c Crimea sits in the Black Sea. It belongs to Ukraine. But Russia annexed it in 2014, and now controls it. (This move was condemned as illegal by the United Nations.)

d Kaliningrad is an exclave of Russia, and lies along the Baltic Sea. (It borders Poland and Lithuania.)

3 Sochi is more southerly than most of Russia, so is one of Russia's warmest places. It is on the coast of the Black Sea – and everyone likes to holiday by the sea. It is near the Caucasus Mountains, so tourists can go trekking in the mountains in summer, and ski in them in winter.

4 The warm ocean current called the North Atlantic Drift keeps it warmer in winter. This current (an extension of the Gulf Stream) keeps the west coasts of the British Isles warmer in winter too.

5 Students could choose three from these.
*Push factors:* few jobs; little choice of work; low income; few services; poor schools; people may be tired of farming.
*Pull factors:* a greater choice of work; better pay; more services; better schools; life more interesting, etc.

6 a St Petersburg and Kaliningrad, Baltic Sea; Murmansk, Arctic Ocean (Barents Sea); Sevastopol, Black Sea.

b Ports allow you to trade easily with other countries – and you can also send out warships and submarines if necessary.

From the ports on the Baltic Sea you can access many European countries which do not share borders with Russia, including Denmark, Germany, the UK, and France. (You also access the rest of the Atlantic Ocean.)

From Murmansk you access the Arctic Ocean (part of which Russia is entitled to exploit). The Arctic Ocean has big reserves of oil and gas, as well as fish stocks. If it becomes more ice-free for more of the year, this gives Russia a quicker route to Europe and North America, as well as to places on its own north coast.

From Sevastopol you can access the Mediterranean Sea through the Turkish straits, and from there, the countries which border the Mediterranean. You can go through the Suez Canal and down the Red Sea to the Indian Ocean. Again this is useful for both trade and strategic purposes.

7 From what they have learned already, students might expect to see: vast expanses of taiga; great rivers (frozen in winter); steppe; some mountain scenery; and exciting volcanic landscapes in the Kamchatka Peninsula and Kuril Islands. They should expect to see fewer settlements.

# 6.6 Climate change in Yakutia

Help at a glance

## About this unit

In this unit students will explore how rapid warming is affecting Yakutia, in Siberia, Russia.

## Key points

- Yakutia, in Siberia, is Russia's coldest region, and sits on permafrost.
- Winters last for eight months with temperatures as low as –40°C.
- The climate in Russia is warming 2.5 times faster than the global average.
- One of the main impacts of the warming is the thawing of permafrost.
- Mining and farming may benefit from the warming climate.

## Key vocabulary

Yakutia, Republic of Sakha, permafrost

## Skills practised in 'Your turn'

- Geographical skills: q3, q8, interpret photos
- Literacy skills: q3, define
- Thinking skills: q1, q2, q3, q4, q6, q7, q8, explain; q3, imagine a scenario; q5, give two examples; q6, identify an impact; q7, describe

## Unit outcomes

By the end of this unit most students should be able to:

- explain the terms given in 'Key vocabulary' above
- explain why buildings in Yakutia are built on concrete stilts, and the impact that permafrost melting is having on buildings
- draw a mind map to show how climate change is affecting Yakutia
- explain how climate change might affect population density and average wealth per person in Yakutia

## Ideas for a starter

1. Ask: Where is Siberia? What do you know about it? How is it different from European Russia?
2. Find, and show students, a map of minimum air temperatures for Russia. The intense cold in Siberia will show up clearly.
3. Ask a student to read out 'Nightmare on Avtodorozhnaya Street' to the rest of the class, to set the scene for the lesson.

## Ideas for plenaries

1. Quick quiz. What is permafrost? Why build houses on stilts? Why does permafrost melt? What's the connection between permafrost and greenhouse gases? Permafrost and mammoth tusks?
2. Use 'Your turn' question 8 as a plenary activity.

## Further class and homework opportunities

Suggestions 18 and 19 on page 147 of this book

*geog.3 workbook*, page 55

kerboodle

### Resources

6.6 You decide! – Students rank the challenges of living with permafrost.

6.6 Quiz time! – Students complete the explanation of climate change in Yakutia.

6.6 Foundation worksheet – Students insert numbers into a paragraph describing Yakutia, create a factfile on the region, and consider how climate change has affected the area.

6.6 Extension worksheet – Students create a factfile on Yakutia, and analyse the effects of climate change on Russia and how this might impact Russia's economy.

### Assessment

6.6 End-of-lesson test – An assignable, auto-marked formative assessment.

## Answers to 'Your turn'

1 Few people live in Yakutia due to the extreme climate. It is Russia's coldest region and over 40% of the land lies within the Arctic Circle. It sits on permafrost and experiences winters which last eight months, with temperatures below –40°C. Due to its latitude, some of its winter months are in complete darkness. Another reason for the low population density is its remoteness. Yakutia is a long way from Moscow, the administrative centre of Russia, and there are not many transport routes in, or out of, the region. For much of the year, the rivers and are sea frozen, so boats can't be used.

2 Yakutia is important due to the resources which are mined there. Among these are diamonds, gold, coal and oil – all of which contribute to Russia's GNI.

3 **a** Permafrost is the ground under the surface that is permanently frozen. It is found in the tundra regions.

**b** Buildings are on stilts in Yakutsk because they are built on permafrost. The stilts keep them stable in the summer when the upper layer of permafrost thaws. Stilts also reduce heat transfer from the building to the ground, which might thaw the permafrost further, making it more unstable.

**c** People living in the building could be injured or even face death if the walls suddenly collapsed due to melting permafrost. More likely, cracks would appear over time and the building would be evacuated and fenced off. The residents would need to find somewhere else to live, which might be difficult because buildings nearby are likely to have suffered similar fates.

4 The south of Yakutia lies south of the Arctic Circle, and therefore temperatures are likely to be warmer than further north (due to the effect of latitude). The Lena will thaw earlier in the south, providing water for crops and animals.

5 The main issue linked to climate change is deeply thawing permafrost, which has made the pastures soggy and lumpy. Animals which are reared there (such as cattle) find it hard to graze. Floods are also becoming more common as rivers overflow during the melting season. Other issues include wildfires and heatwaves.

6 Students are likely to mention rising temperatures caused by the increase of greenhouse gases in the atmosphere. This leads to a variety of issues around the world such as: rising sea levels (due to melting ice caps); droughts and floods becoming more common (due to changing precipitation patterns); pests and diseases spreading; loss of biodiversity as wild plants and animals struggle to adapt. People are worried because these issues cause suffering to humans and are damaging Earth's natural balance.

7 **a** In Yakutia, population density might decrease if people move to other parts of Russia which are not being affected by melting permafrost. But some farmers might move to the region's capital Yakutsk, where population density may increase.

**b** Climate change could affect average wealth per person in a few ways. Firstly, damage to buildings and infrastructure would be difficult to repair and solutions may be expensive. Secondly, income from farming would be lost and food may have to be imported at a high cost. Thirdly, if the world cuts back on the use of fossil fuels, Russia will lose income from the sale of oil and gas. (On the other hand, if ports are free of ice during winter months, Russia would be able to increase trade and mining in the Arctic Ocean, and land underneath the ice could be farmed).

8 Answers will vary.

# 6.7 What is Russia doing in the Arctic Ocean?

Help at a glance

## About this unit

This unit consider how Russia is responding to the fact that the Arctic Ocean is likely to become ice-free in summer.

## Key points

- Five countries in the Arctic region have territory within the Arctic Circle.
- Experts predict that the Arctic Ocean will be ice-free in summer by 2040.
- Russia has the longest coastline on the Arctic Ocean, and sees opportunities to exploit oil, gas, metal ores, fish, tourism, and the Northern Sea Route, as the summer ice melts.
- Russia is also building military bases along the coast, expanding its ice-breaker fleet, inviting foreign countries and investment to extract resources and develop infrastructure.
- Russia's plans risk increasing greenhouse gas emissions, and damaging the ocean and its ecosystems.

## Key vocabulary

Arctic Ocean, Arctic region, exclusive economic zone (EEZ), Northern Sea Route, icebreakers, Belt and Road initiative

## Skills practised in 'Your turn'

- Geographical skills: q1, q2, q5–q6, interpret maps; q6, compare maps
- Literacy skills: q4, define; q8, write a balanced argument
- Thinking skills: q1, make a link; q3, q4, q6, explain; q5, comment; q7, suggest; q8, consider an opinion

## Unit outcomes

By the end of this unit most students should be able to:

- explain the terms given in 'Key vocabulary' above
- name the countries that have coastlines on the Arctic Ocean
- explain how, and why, the summer ice cover of the Arctic Ocean is changing
- give three ways that Russia can exploit an ice-free Arctic Ocean
- write a list of points for and against exploiting the Arctic Ocean

## Ideas for a starter

1. Show students the photo of the polar bear on page 120 of the students' book. Ask: What is the bear doing? Why?
2. Who can remember what the Belt and Road Initiative (BRI) is? (It's a project to create a network of links from mainland China across the world, consisting of roads, railways, pipelines, ports, etc. Unit 8.8 in *geog.2* looks at the BRI.)

## Ideas for plenaries

1. Play 'Pass the parcel' to Russian music. Prepare a box with questions to cover all aspects of the work done on Russia – some true, some false. Each time the music stops, the person holding the box takes a question and answers it. If correct the question is discarded, if incorrect the question is returned to the box.
2. Ask students to create a mind map to sum up their work on Russia. Give them a time limit.

## Further class and homework opportunities

Suggestions 20 and 21 on page 147 of this book

*geog.3 workbook*, page 56

**kerboodle**

### Resources

6.7 You decide! – Students rank the benefits of the melting Arctic sea ice.

6.7 Word scramble – Students complete a crossword of key terms about Russia.

6.7 Foundation worksheet – Students list the benefits, advantages and disadvantages of Russia having an exclusive economic zone (EEZ) in the Arctic.

6.7 Extension worksheet – Students explain how the Arctic ice melting might increase Russia's GDP and consider the global impact of Russia's exploiting the Arctic Ocean.

### Assessment

6.7 End-of-lesson test – An assignable, auto-marked formative assessment.

## Answers to 'Your turn'

**1** Greenland is a territory of Denmark, which means it is part of the Danish administration. It was a Danish colony until 1953, when it was redefined as a district of Denmark. It has its own government and its own parliament, but its head of state is the Danish monarch.

**2** Russia, USA, Canada, Denmark, Norway.

**3** In summer, much of the sea ice melts in the Arctic Ocean. Climate change is causing more of it to melt every year. This is because sea temperatures are increasing, which leads to more melting. (Some students may also know about the albedo effect. As more ice melts, there are fewer white surfaces in the Arctic to reflect the sun's energy, so more heat energy is absorbed, causing further melting).

**4** **a** An exclusive economic zone (EEZ) is an oceanic area which a country has rights over. This can be for fishing, mining or other industry.

**b** Russia has the longest coastline in the Arctic Ocean, so its EEZ is more extensive.

**5** Students should notice that Russia is making claim to two areas of the Arctic Ocean which are further than 370 km from its coastline and are not part of its official EEZ. The largest area includes the North Pole. Since these areas are not near the coastline of other countries, students can assume that Russia is not encroaching on their territory. However, making this claim indicates that Russia wants to expand its economic and political dominance.

**6** **a** China is interested in an ice-free northern sea route so that it can trade more easily with Europe. This route is much shorter than the alternative, which passes around Asia, through the Suez Canal and along the coast of north Africa to get to Europe. It may also be worth mentioning that the Suez Canal is a narrow 'chokepoint,' and students may recall the blockage which occurred when a ship got stuck in March 2021. An ice-free Arctic Ocean would also make trade between China and Russia easier. If students look at the map on page 140–141, they may notice that China is close to Russia and assume that goods could be transported overland – but they should consider that mountain ranges may affect these routes.

**b** An icebreaker is a ship that forces its way through sea ice, clearing a path for other vessels.

**7** Russia could exploit fossil fuel and mineral reserves by mining; it could expand its fishing capacity; it could increase tourism – particularly excursions to the North Pole.

**8** Answers will vary but students should consider the economic benefits of resource extraction vs the environmental benefits of protecting species that rely on this ocean, and the global impacts which may result from damage to Arctic food chains and as a result of oil and gas extraction.

# About Russia



## Skills practised in check questions

- Geographical skills: q1, q2, q3, interpret maps; q2, q4, interpret a photo; q3, interpret a table; q5, interpret a pie chart
- Literacy skills: q2, q4, define; q6, extended writing
- Numeracy skills: q2, estimate percentage; q3, divide
- Thinking skills: q2, q4, explain; q2, q3, q4, suggest impacts; q3, q4, q5, give reasons; q5, give evidence; q6, weigh up

## Answers for check questions

**1 a i** Arctic Ocean **ii** Sea of Okhotsk

**b i** Ural Mountains

**ii** Asian Russia / West Siberian Plain

**c i** Lena **ii** Tundra and taiga

**d i** St. Petersburg and Moscow

**ii** 6. Moscow

**2 a** The ground under the surface that is permanently frozen, in the tundra

**b** The answer is **ii** About 65%

**c i** The thaw is deeper every year because climate change is causing rising temperatures in permafrost regions.

**ii** One from: farmers are being displaced by changing conditions; buildings are being damaged as their foundations shift; infrastructure such as pipelines may be damaged; it is becoming dangerous to use frozen rivers as roads during winter; more floods; more wildfires; more heatwaves.

**iii** One from: people may be able to grow more crops as melting permafrost reveals fertile farmland; more money can be made from mining; woolly mammoth tusks found under the permafrost can be sold.

**iv** The impacts of climate change are global. In the UK, temperatures are increasing and climate patterns are becoming more unstable; the summer has more frequent heatwaves and storms; mosquitoes and other pests are now more common in southern parts of England; rising sea levels are affecting coastal areas – billions have been spent protecting London and the Thames Estuary. The UK has links with other countries and further impacts may be created by the challenges they face: higher immigration rates; food shortages; rising global poverty levels.

**3 a** Yakutia is a remote region with an extreme climate, and rests on permafrost. By contrast, Russia's central region has a milder climate, more fertile farmland, and is closer to Europe which makes trade easier.

**b** More people may move to Yakutia if the climate warms up. This may take some time for the vast majority of the region, because thawing permafrost has made the ground unstable and unsafe for buildings. However, if ice melts in coastal towns, ports may open up and people may move there to work in shipping, fishing or offshore mining.

**c** About 30 times greater

**d** One from: some Russians live thousands of miles from the capital; people in Russia speak many different languages; there are cultural differences between regions; people may not agree on laws and policies that suit everyone.

**4 a** The replacement fertility rate is 2.1 babies per woman. Lower rates will lead to a fall in population (unless immigration rates make up the difference).

**b** The population will be imbalanced if less babies are born. There will be too many older people that require support from the younger people that are working; this may require a rise in taxes. If the population falls dramatically, there may not be enough people to do the work required to keep the country's economy afloat.

**c i** Inequality occurs when wealth and access to services are not shared equally.

**ii** In 1991, the USSR broke apart and republics became independent countries. Collapse of a political system usually leads to poverty until governments stabilise. In Russia's case, oligarchs bought state-owned industries in 1991 and took the profits for themselves.

**iii** Inequality can create extreme poverty for the poorest in society. These people may struggle to feed, house and clothe themselves. They may suffer health issues and some may be unable to work. Political unrest may result; or foreign aid may be required to support these groups.

**5 a** The pie chart on page 109 shows that in 2020, oil and oil products made up over 60% of Russia's exports. Gas and coal made up another 15%. The term *exports* is the main clue here, as it implies these are being sold to other countries. At 75% of total exports, this is strong evidence that Russia's economy relies heavily upon selling fossil fuels.

**b** The data on the pie chart was from 2020 but recently, there has been a global push to reduce the use of fossil fuels to tackle the challenge of climate change, so this percentage may change. To keep up, Russia may need to start producing green energy or focus on other exports and / or services.

**6** For this extension question, students should consider whether benefits outweigh costs. Good answers will refer both to Russia and to the rest of the world and provide a short conclusion.

An example answer: One benefit of climate change in Russia is that melting sea ice has opened up the northern sea route for more months of the year, providing a shorter path for ships travelling to Europe. This could benefit both Russia and Europe, as trade will be easier and more profitable. If ships have better access to the Arctic Ocean, there may be more opportunity for Russia to mine for fossil fuels, which would boost its economy; but this is likely to impact negatively on global climate in the long term.

Another potential benefit is that melting permafrost may create ideal conditions for agriculture in Siberia. Russia could grow more food, which it could sell.

Ultimately, the economic benefits might outweigh the damage done to Russia by climate change in the short term; but in the long term, and for the rest of the world, the environmental costs are likely to be much more significant than any of the benefits.

# Further suggestions for class and homework

Most of these suggestions are addressed to your students. Where research or further resources are needed, the Internet will almost certainly provide the answer.

The suggestions are graded *, **, *** according to level of difficulty. Some are suitable for all levels, and can be differentiated by outcome.

## Meet Russia!

1 **Russian exposure** Note the names of Russian places and people mentioned in the news / newspapers, daily for at least two weeks. Mark places on a map of Russia, with a tally mark. Keep a tally for people. Is there a pattern? You could allocate this as a rolling task to different students. **/***

2 **Russian recipe** See if you can find a Russian recipe you like the look of. Then follow it! Share with the class? **/***

3 **Wall display** Start planning for a big wall display for Russia. What kinds of things will you show? Don't forget currency, and Russian script. Students could work in teams. */**/***

4 **Alaska, a Russian colony** How did the Russians get there? What did they do there? How much did they sell it for? ***

## What are Russia's main physical features?

5 **Russia from space** Download a satellite image of Russia – for example from Google Earth or Google Maps. Print it out, paste it onto a larger sheet of paper, and annotate it with the names of the physical features you can identify. */**/***

6 **Snap** Make a set of cards about physical features of Russia. Use two cards for each feature, writing part of the name on the first card, and the rest on the second. For example *Lake* on one card and *Baikal* on the second, or *Kamchatka* on one card and *Peninsula* on the second. Then devise a way to play a game like Snap. **/***

7 **Go sailing** On an outline map of the world, mark in the shortest routes ships could take from Russia to the UK, Italy, China, North America! Mark in Russian ports they could leave from, and the seas and oceans they'll travel across. ***

8 **Be map ready!** Practise drawing an outline map of Russia to help you create a good mental map. Then step up a few gears, and mark in its neighbours too! */**/***

## Russia's climate zones and biomes

9 **Compare climate graphs** Choose places in different climate zones in Russia. Find climate graphs for them on the Internet, showing average temperature and precipitation each month. For example, Moscow, Noril'sk, Sochi. Make up questions for a partner to answer. For example: Which month is wettest / driest / hottest / coldest in Moscow? Which of these place gets hottest? Why? Which one gets coldest? Why? (You may need to select the graphs for students. They could work in teams, each choosing a different combination of places.) **/***

10 **Tundra** Write a unit about the Russian tundra, like the units in *geog.3*. Describe the climate and landscape. Do any indigenous groups live there? How do they live? */**/***

11 **Permafrost** Agree on a good enquiry question about Russia's permafrost – and then carry out the enquiry. (They'll find photos, and a map of extent, on the Internet.) */**/***

## What about the Russian people?

12 **A famous Russian** Choose one to find out about. Tell us about them. */**/***

13 **Old Peter's Russian Tales** Traditional tales collected by Arthur Ransome. Choose one and read it to the class? (For a younger age group, but gives a flavour of Old Russia.) */**/***

14 **Moscow** How many people live there? What's it like? What is life like for an ordinary Russian in Moscow? Find out and tell us. A podcast? */**/***

## A tour of European Russia

15 **Plan a school trip ...** for Year 9 to European Russia. Away for eight nights, flying into St Petersburg, and from there by train to other places. Which month? Write an itinerary. Name at least one attraction you'll look at in each place. Draw a map of the itinerary. Suggest clothing to bring. **/***

16 **Visit Sochi ...** on Google Earth. What signs can you see that it's a seaside resort? And that it suffers from beach erosion? You can save, download, and annotate images. */**/***

17 **Visit Russia!** Create a colourful poster and slogan, to attract tourists. It will appear on billboards. */**/***

## Climate change in Yakutia

18 **Yakutsk photos** Find photos showing Yakutsk in summer and winter. **/***

19 **A letter from Yakutia** You live in Yakutia, and you're worried – about climate change, the impacts it will have and what it means for your future. Write a letter – include some drawings or photos. */**/***

## What is Russia doing in the Arctic Ocean?

20 **Update!** It's now ... (fill in the date and year). What has happened in the Arctic Ocean? How much ice melted last summer? What has Russia been doing in the Arctic since this unit was written (in 2021)? Find out. **/***

21 **An Arctic refuge** Plan a refuge in the Arctic region. No extraction of oil, gas or minerals allowed. Only sustainable tourism. Wildlife protected. What else? make a poster. */**/***

# 7 The Middle East

## About this chapter

### Addressing the KS3 Programme of Study

- Extend locational knowledge and deepen spatial awareness of the world's countries, using maps of the world to focus on … the Middle East, its environmental regions, key physical and human characteristics, countries, and major cities

### About the content

- We take the Middle East to consist of 16 countries, and the State of Palestine. Only one is a North African country: Egypt.
- This chapter looks at the physical and human geography of the Middle East and the challenges the Arabian Peninsula faces.
- One unit takes a broad look at conflicts in the Middle East. These are fluid and changing, but we hope this unit will provide useful background.

## The big picture

These are the key ideas in this chapter.

- The Middle East is made up of 16 countries and the State of Palestine.
- The Middle East has always been a strategically important region, sitting as it does where Africa, Europe, and Asia meet, between the Mediterranean Sea and Indian Ocean.
- Today it is economically important too, because of its large reserves of oil and gas, which have made several Middle East countries very wealthy. However, countries are now diversifying as the world moves away from fossil fuels.
- The majority of people in the Middle East – but not all – share Arabic as their main language, and Islam as their religion.
- Overall, the region is hot and dry. But it does have different climate zones, and corresponding biomes.
- The Arabian Peninsula faces a number of challenges, including water stress, food risk, a rapidly growing population, climate change, and a fall in demand for oil and gas.
- The region has long been unstable, suffering from conflict.

## The chapter outline

Use this to give students a mental roadmap for the chapter.

**7** **The Middle East** The chapter-opening photos in the students' book are an important resource. See opposite for using them.

**7.1** **Introducing the Middle East** Where it is, the countries and state which comprise it, an overview of the region, and its links with Britain

**7.2** **The Middle East: physical geography** A map and look at its main physical features

**7.3** **The Middle East: climate zones and biomes** Describes and explains the climates, and introduces the corresponding biomes

**7.4** **The people, and economies** Population, exports, and wealth

**7.5** **Challenges in the Arabian Peninsula** How the Arabian Peninsula is tackling some of the challenges it faces

**7.6** **Conflict in the Middle East** Background to the conflicts in the region

**7** **The Middle East check questions** A page of questions to check knowledge, understanding and skills across the chapter, with an opportunity for extended writing

## Objectives and outcomes for this chapter

| Objectives | Unit | Outcomes |
|---|---|---|
| Most students will understand: | | Most students will be able to: |
| • what and where the Middle East is | 7.1, 7.4, 7.5 | • locate the Middle East and the Arabian Peninsula on a world map; name at least 12 of the countries and their capitals (including all in the Arabian Peninsula) |
| • that it is almost surrounded by bodies of water; that some of it is mountainous, but much is flat; that it has three great rivers | 7.1, 7.2 | • name and locate the bodies of water which border the region also the Suez Canal and Strait of Hormuz; describe the pattern of relief; name and locate at least three mountain ranges; name and locate the Nile, Euphrates, and Tigris rivers; explain the significance of a choke point |
| • that overall the region is hot and dry – but with distinct climate zones, and corresponding biomes | 7.3 | • explain why the region is hot and dry; describe and roughly locate the climate zones and biomes |
| • that much of the region is sparsely populated and that it contains many ethnic groups of which the largest is Arab; that wealth varies across the Middle East | 7.4 | • describe and explain the pattern of population density; name the four largest ethnic groups by area; describe how wealth varies across the region |
| • that the Arabian Peninsula faces a number of challenges | 7.5 | • list the challenges facing the Arabian Peninsula; give examples of how some of the challenges are being tackled |
| • that the region is unstable, with conflict | 7.6 | • give at least three causes of conflict (past and present); outline why foreign countries get involved in the Middle East; give four examples of conflict, in the region |

These tie in with the outcomes for each unit in this teacher's handbook.

## Assessment for this chapter

There are assessment materials for this chapter on *geog.3 Kerboodle Lessons, Resources, & Assessment*. They include interactive auto-marked end-of-lesson and chapter-level tests, a printable exam-style end-of-chapter paper (with a mark scheme), and a printable self-assessment checklist. See also the notes on pages 14–17 of this book.

## Getting ready for this chapter

There is a scheme of work for this chapter, plus a glossary and glossary worksheet, on *geog.3 Kerboodle Lessons, Resources, & Assessment*.

## Using the chapter-opening photos

Ask students:

- These photos are all from the Middle East. Which continent do you think the Middle East is in? Are you sure?
- What can you say about the Middle East from these images?
- What else do you know about the Middle East?

## About 'Your chapter starter'

The photos on page 123 of the students' book show, row by row and left to right: the Burj Al Arab hotel in Dubai, United Arab Emirates, built on an artificial island; young shepherds in Turkey; skiing in Turkey; a satellite image focusing on the Middle East (with parts of other regions showing); camels in the desert in Saudi Arabia; saffron from Iran, and other spices, on sale in the spice market in Istanbul, Turkey; young students in Saudi Arabia

# 7.1 Introducing the Middle East

Help at a glance

## About this unit

This unit introduces the Middle East: where and what it is.

## Key points

- The Middle East sits where Africa, Europe, and Asia meet: a strategically important location.
- It consists of 16 countries, and the State of Palestine.
- Most of it is desert, or dry land, but is also has mountain ranges, forests, fertile farmland, and beaches.
- The region has large oil and gas reserves which have made some countries wealthy.
- The region has a lot of conflict and poverty (which is linked to conflict).
- The Middle East has a rich mix of culture and ethnic groups.
- Britain has been involved in the Middle East since 1882 which led to unrest, and eventual independence.

## Key vocabulary

United Arab Emirates, Arabian Peninsula, Allies

## Skills practised in 'Your turn'

- Geographical skills: q1, q2, q4, q5, interpret maps
- Literacy skills: q6, list key points
- Thinking skills: q1, unscramble country names; q3, define; q6, imagine a scenario

## Unit outcomes

By the end of this unit most students should be able to:

- explain the terms given in 'Key vocabulary' above
- show the region on a world map, and name at least 12 of its countries
- describe at least four key points of the Middle East region

## Ideas for a starter

1. Ask: What's the Middle East? Tell me anything you know about it. Write responses on the board. You can come back to them later.
2. Show varied images from the Middle East. Include fertile land, desert, modern cities, wealth, poverty, conflict. What impressions do they give? Write adjectives on the board.

## Ideas for plenaries

1. Look at the Middle East region on map A. What continents does it lie in? How would you describe its shape?
2. Without looking at other maps, see if you can name any of those bodies of water around the Middle East. (Answers are on the map on page 126 of the students' book.)
3. They say the Middle East is in a strategic location. Why do people think that?
4. Start a big spider map for the Middle East, on paper. Which headings do you think you should use? Students could work in pairs.
5. If you chose starter 1, you could return to the board. What should be added / deleted?

## Further class and homework opportunities

Suggestions 1–4 on page 163 of this book

*geog.3 workbook*, page 57

kerboodle

### Resources

7.1 Word scramble – Students complete a crossword of countries in the Middle East.

7.1 Quiz time! –Students complete a quick quiz about the Middle East.

7.1 Foundation worksheet – Students fill in gaps in a paragraph about the Middle East, describe the location of three countries, and produce a map of the region from memory.

7.1 Extension worksheet – Students fill in gaps in a paragraph describing the Middle East, outline the Ottoman Empire's decline, and explain the location and importance of key areas.

### Assessment

7.1 End-of-lesson test – An assignable, auto-marked formative assessment.

## Answers to 'Your turn'

**1** **a** Turkey **b** Syria **c** Iran **d** Qatar **e** Lebanon **f** Egypt **g** Yemen **h** Oman

**2**
- **a** Saudi Arabia (about 2.1 million sq km)
- **b** Iran (about 1.6 million sq km)
- **c** Bahrain
- **d** Bahrain and Cyprus
- **e** Turkey, Syria, Jordan, Saudi Arabia, Kuwait, Iran
- **f** Turkey, Lebanon, Israel, Jordan, Iraq

**3** A boot-shaped peninsula in the Middle East which borders the Red Sea, and juts into the Indian Ocean. It contains Saudi Arabia and five other, smaller countries. A seventh country, Bahrain, lies off its east coast, and is connected to Saudi Arabia by a causeway.

**4** Students should notice these suspiciously straight borders:
- between Egypt and its neighbours
- between Jordan and its neighbours
- between Saudi Arabia and its neighbours
- between Yemen and Oman.

Note that the current borders were not all established by Britain and France. Some border discussions went on for decades.

**5** Any four of: Egypt, the land which is now Israel and the State of Palestine, Jordan, Iraq, Qatar, UAE, Oman, Yemen. (For information: Egypt was a British Protectorate from 1882 to 1953.)

**6** Answers will vary.

# The Middle East: physical geography

Help at a glance

## About this unit

This unit focuses on the main physical features of the Middle East.

## Key points

- Iran and Turkey are the two most mountainous countries of the Middle East. There are also mountains down the west coast of the Arabian Peninsula, and at its heel and toe.
- The region has many rivers including the Nile, Tigris, and Euphrates.
- Most of Egypt and the Arabian Peninsula are desert.
- Bodies of water around it give the Middle East great strategic importance. There are three maritime choke points in the Middle East.
- Several tectonic plates meet in the region. So Turkey and Iran are prone to earthquakes. The Arabian plate is moving away from the African plate, so the Red Sea is slowly growing wider, and contains active volcanoes.

## Key vocabulary

maritime choke point

## Skills practised in 'Your turn'

- Geographical skills: q1–q5 interpret maps; q7, interpret a table
- Thinking skills: q6, describe; q5, q7, explain

## Unit outcomes

By the end of this unit most students should be able to:

- explain the term given in 'Key vocabulary' above
- name and locate the physical features mentioned in this unit
- explain the importance of the Suez Canal, the Bab el Mandeb Strait, and the Strait of Hormuz
- explain why Turkey and Iran have earthquakes, and the Red Sea has volcanoes

## Ideas for a starter

1. With books closed, say: We are going to study the physical geography of the Middle East. What is *physical geography?*
2. With books closed, say: You will find all of these in the Middle East. Who can unjumble?

| | | | | | | | |
|---|---|---|---|---|---|---|---|
| *rveris* | *sfgul* | *esdrset* | *mtaunnsio* | *coaneslvo* | *ttariss* | *aess* | *oaess* |
| (rivers | gulfs | deserts | mountains | volcanoes | straits | seas | oases) |

## Ideas for plenaries

1. Look at the photos on pages 126 and 127 of your books. Why is there snow on Mt Damavand? Is the Suez Canal a natural feature? How was a new island created in the Red Sea in 2012?
2. What is a strait? (a narrow waterway connecting two larger areas of water) A gulf? (an area of ocean partly enclosed by land) A sea? (a stretch of salt water connected to, or part of, the ocean, with a shape that's defined by the land)
3. With books closed, write up answers like those following. What is the question?

   Strait of Hormuz; The Red Sea and the Mediterranean Sea; No, it flows through several other countries too; They're called the Zagros Mountains; Because nobody lives there – it's just sand.

## Further class and homework opportunities

Suggestions 5–8 on page 163 of this book

*geog.3 workbook*, page 58

**kerboodle**

### Resources

7.2 Where in the world? – Students match the physical features to the Middle Eastern country.

7.2 Quiz time! – Students complete a description of the Middle East's physical geography.

7.2 Foundation worksheet – Students analyse tectonic activity in the Middle East, explain the importance of the Suez Canal, and design a word search using five key terms.

7.2 Extension worksheet – Students explain why natural disasters threaten the Middle East, sketch and colour a map of the Middle East, and design a word search using ten key terms.

### Assessment

7.2 End-of-lesson test – An assignable, auto-marked formative assessment.

## Answers to 'Your turn'

**1** **a** Three from: Turkey, Iran, Yemen, Lebanon

**b** **i** Iran **ii** Saudi Arabia **iii** Turkey

**c** Mt. Damavand in Iran

**2** **a** Clockwise from the Arabian Sea: Arabian Sea (which is part of the Indian Ocean), Gulf of Aden, Red Sea, Mediterranean Sea, Black Sea, Caspian Sea, Arabian Gulf, Strait of Hormuz, Gulf of Oman. (Note that a strait is a narrow passageway of water.)

**b** the Arabian Gulf and Gulf of Oman

**3** The Indian Ocean

**4** **a** Nile, Euphrates, Tigris

**b** Because there's so little rain (but some rivers do appear temporarily, after rain)

**5** **a** Students should write a paragraph about the Suez Canal including its location (Egypt) and other facts such as: it is 193 km long, it is 205 m wide, it is used by about 50 ships a day, ships pay to use it because it is a short cut which saves them travelling around the whole continent of Africa.

**6** It is a large continuous stretch of sand in the south of Saudi Arabia, practically uninhabited – and note that it contains vast oil reserves.

**7** **a** The boundary between the African and Arabian plates lies along the Red Sea – and the Arabian plate is moving away from the African plate. That is why the Red Sea exists in the first place, and why it is getting wider.

**b** Turkey is in a highly-active tectonic zone. It lies on two small plates (the Anatolian plate on the right, and Aegean Sea plate) which sit between the Eurasian, African, and Arabian plates, and are being pushed westwards. The enormous pressure has produced many faults in Turkey, along which earthquakes occur.

**c** Yemen

# The Middle East: climate zones and biomes



## About this unit

This unit investigates the climate zones and biomes of the Middle East.

## Key points

- Generally, the Middle East has a hot, dry climate.
  - It is hot because of its latitude; it lies partly in the tropics.
  - It is dry because dry air descends in the region, as part of a cell of circulating air (the Hadley cell). As dry air descends it warms, so any moisture it does contain will not condense, so no rain forms.
- Although overall hot and dry, the region does have distinct climate zones: arid, semi-arid, and Mediterranean.
- The different climates lead to different biomes.

## Key vocabulary

arid, semi-arid, Mediterranean, biome, desert, steppe

## Skills practised in 'Your turn'

- Geographical skills: q1, interpret a diagram; q2, q3, interpret maps
- Literacy skills: q1, fill in words to complete a paragraph
- Thinking skills: q2, q3, explain; q4, choose and explain choice

## Unit outcomes

By the end of this unit most students should be able to:

- explain the terms given in 'Key vocabulary' above
- explain why the Middle East is hot and dry, overall
- describe and roughly locate the region's climate zones
- describe and roughly locate the region's biomes

## Ideas for a starter

**1** With books closed, write *Climate* on the board. Then ask: What does *climate* mean? What aspects does it have? (Temperature, rainfall, wind speed, cloud cover, hours of sunshine, and so on.) Which aspects are we usually the most interested in?

**2** Find and display a climate graph for Kuwait City (with a single temperature line). Ask questions about it. For example: Which month is coldest? hottest? Is there rain every month? Then do the same for Sanaa in Yemen. Is there much difference? Can you explain why? (Latitude is a factor. So is altitude: Kuwait City, 10 m, Sanaa, 2259 m.)

## Ideas for plenaries

**1** Make sure students understand diagram A. In the Middle East, for example, Kuwait is 30 °N. Thanks to this circulation pattern, dry air will descend there, so there'll be little rain. (You could show the climate graph for Kuwait City.) Then the dry air is pulled south. Say that this circulation cell has a major effect on rainfall – but other factors still play some part. For example altitude (when air is forced to rise and cool, and any moisture in it condenses), and distance from the sea.

**2** Map B gives a clue to a very big crisis facing the Middle East region. What do you think it is? (A shortage of water.) Then turn to the map on page 27 of *geog.3*. Point out that the Middle East has physical water scarcity. That issue is followed up in suggestion 8 on page 63 of this book.

## Further class and homework opportunities

Suggestions 9–11 on page 163 of this book

*geog.3 workbook*, page 59

**kerboodle**

### Resources

7.3 Sort it out! – Students complete a quick quiz about the Middle East's climate and biomes.

7.3 Quiz time! – Students complete the explanation of why the Middle East is mainly desert.

7.3 Foundation worksheet – Students produce three multiple choice questions about the climate in the Middle East, draw and label a biome, and evaluate tourist destinations in the region.

7.3 Extension worksheet – Students produce five multiple choice questions about the climate in the Middle East, sketch and annotate a biome, and draw a diagram explaining Hadley Cells.

### Assessment

7.3 End-of-lesson test – An assignable, auto-marked formative assessment.

## Answers to 'Your turn'

**1** The Middle East is *hot overall*, because of its *latitude*. It is also dry, because it lies where dry air descends, after losing *moisture* over the *Equator*. Because most places have little rain, vegetation tends to be *sparse*. But some places have enough rain for *forests* to grow.

**2** **a** There are warm, wet winters in the Mediterranean zones in the north (Cyprus, Lebanon, coastal regions of Turkey, coast of Syria, north and north-west Iran).

**b** Because there is not much rain. Most of the Middle East is dry or very dry. (But some countries lie over aquifers of ancient fresh water, or *fossil water*, and are pumping it up, and using it up fast. There is not nearly enough rain to recharge the aquifers.)

**3** **a** There are no natural forests in Oman because it is mostly arid desert, which is very dry. The only vegetation which can survive in these conditions is tough and spiky, and it is also sparse. In the north of Oman, there is a narrow coastal strip of semi-arid land, but even this can't support forests. Common plants in the semi-arid climate zone are grasses and low bushes.

**b** Saudi Arabia has to import most of its food because, like Oman, it is mostly desert. Deserts have poor soil and not enough rain to support agriculture. Growing food in the desert is possible, but only with irrigation systems, which are expensive and require vast amounts water (see photos on pages 125 and 133 of the students' book).

**4** Answers will vary.

# 7.4 The people, and economies

Help at a glance

## About this unit

This unit looks at the people of the Middle East, exports, and wealth.

## Key points

- Population density in the Middle East varies, influenced by climate.
- The region has many different ethnic groups, Arabs make up the largest group.
- Oil and gas are the Middle East's main exports, but cars, clothing, high tech equipment, pharmaceuticals, and agricultural exports are also important for some countries.
- Average wealth per person varies across the region – people are wealthiest in Qatar, and poorest in Syria.

## Key vocabulary

population density, ethnic group, Arab, Turk, Persian, Kurd

## Skills practised in 'Your turn'

- Geographical skills: q1, q2, q4, interpret maps; q3, q4, interpret a table; q6, q7, interpret a graph
- Numeracy skills: q7, manipulate data
- Literacy skills: q1, draw and complete a table; q3, use a glossary
- Thinking skills: q2, describe; q2, q5, explain; q2, q7, give evidence; q2, q5, suggest a reason; q7, discuss

## Unit outcomes

By the end of this unit most students should be able to:

- explain the terms given in 'Key vocabulary' above
- describe and explain the pattern of population density in the Middle East
- name the capital cities of at least 12 countries of the Middle East
- list the four largest ethnic groups by area and give their language
- name the top three countries for each of oil and gas reserves
- discuss why there is inequality in wealth across the Middle East

## Ideas for a starter

1. Revision of climate and relief: draw an outline map of the Middle East on the board. Ask students to colour in the climate zones, and fill in a key, and then mark in mountainous areas, using little mountain symbols. Ask: Where do you think people would want to live?
2. Play traditional music from any country in the Middle East.
3. Rapidly display the flags of Middle East countries. Students guess the country.

## Ideas for plenaries

1. Ask questions to make sure students understand map B. Then compare map B with the climate map on page 128 of their books (or with the map on the board from starter 1). Do you notice any similar patterns? Can you explain them?
2. Do you think map B gives an accurate picture of population density? (It is highly simplified. More colour bands would give a the better picture – but then the map starts to get too complex. And note that there are thousands of settlements not shown.)
3. Ask students to vote for a city of the Middle East, to explore on Google Earth now.

## Further class and homework opportunities

Suggestions 12–15 on pages 163 of this book

*geog.3 workbook*, page 60

**kerboodle**

### Resources

7.4 Where in the world? –Students complete activities to check locational knowledge of the Middle East.

7.4 Sort it out! – Students order the countries of the Middle East by population size.

7.4 Foundation worksheet – Students use resources on pages 130 and 131 of *geog.3* to fill in the missing squares of a table on Middle Eastern countries, their populations, and economies.

7.4 Extension worksheet – Students use resources provided on pages 130 and 131 of *geog.3* to complete a table on Middle Eastern countries, their populations, and economies.

### Assessment

7.4 End-of-lesson test – An assignable, auto-marked formative assessment.

## Answers to 'Your turn'

**1**

| Country | Capital city | Country | Capital city |
|---|---|---|---|
| Bahrain | Manama | Lebanon | Beirut |
| Cyprus | Nicosia | Oman | Muscat |
| Egypt | Cairo | Qatar | Doha |
| Iran | Tehran | Saudi Arabia | Riyadh |
| Iraq | Baghdad | Syria | Damascus |
| Israel | Tel Aviv | Turkey | Ankara |
| Jordan | Amman | UAE | Abu Dhabi |
| Kuwait | Kuwait City | Yemen | Sanaa |

**2** **a** Students should notice that:

- the most densely populated area is along the River Nile in Egypt. That's because people depend on the Nile for their water, including water for irrigating crops. The rest of Egypt is desert.
- overall, the population density is higher in the northern half of the region, where it is cooler (further from the Equator) and less arid.
- the population density tends to be lowest in the most arid areas, since water supply will be a problem.
- population density tends to be higher in coastal areas. These are cooler in hot weather, and give access to the sea for fishing and trading.

**b** False. Notice the population density in the Taurus and Pontic mountains in Turkey, the Elburz in Iran, and the mountainous parts of Saudi Arabia and Yemen. (See the map on page 126.)

**c** X is in the arid biome, which is in the Rub' Al Khali – a large, continuous stretch of sand. Few people live in this desert because it is remote, and the hot, dry climate makes it hard to grow food. By contrast, Y is in the Mediterranean biome, which can support more vegetation and has less extreme temperatures.

**3** **a** Egypt, Iran, Turkey

**b** **i** Egypt, Iran, Turkey

**ii** Kuwait, Oman, Lebanon; the State of Palestine (not yet officially a country)

**iii** Three from: Lebanon, Oman, Kuwait, Qatar, Bahrain, Cyprus

**4** **a** Arabs – Arabic; Persians – Persian; Turks – Turkish; Kurds – Kurdish

**b** Five from: Egypt, Lebanon, Syria, Jordan, Iraq, Kuwait, Saudi Arabia, Bahrain, Qatar, UAE, Oman, Yemen

**5** Exports of oil and gas dominate the region because these are valuable commodities and the Middle East has large reserves of both. Many of the processed exports also contain oil- and gas-related products, such as plastics and fertilisers. However, some countries export other products such as textiles, high-tech equipment and pharmaceuticals.

**6** **a** **i** Qatar

**ii** 2.9 million people (in 2020)

**b** **i** People in Syria have low incomes and many live in extreme poverty.

**ii** 17.5 million

**c** Syria has been engaged in a civil war

**d** Qatar, UAE, Kuwait

**7** D shows high levels of inequality in the Middle East. Some countries are very wealthy and others are extremely poor. In the richest countries, such as Qatar and the UAE, many people live lavish lifestyles, whilst others starve in Syria and Yemen. $91 000 per year is the average income in Qatar, which is equivalent to about £65 000. Syria's estimated $1000 GNI is equivalent to about £720 per year. It is important that students know these are average figures for income and that countries also experience internal inequality in wealth. They could discuss inequality in wages in the UK as a comparison.

# 7.5 Challenges in the Arabian Peninsula

Help at a glance

## About this unit

In this unit, students will find out about a range of challenges that the Arabian Peninsula faces.

## Key points

- The Arabian Peninsula consists of seven countries, which have all – with the exception of Yemen – developed rapidly since oil was first discovered there.
- It now faces a number of challenges including: water stress, food risk, a rapidly growing population, a fall in demand for oil and gas, and climate change.
- To tackle water stress, sea water is desalinated, waste water is cleaned and used for farming, the UAE carries out cloud-seeding, and the cost of water might increase.
- To tackle the food issue, in Saudi Arabia the government gave grants to turn desert into farmland – but it's very expensive, so they have invested in farmland overseas and will import crops.
- Plants grown in greenhouses using hydroponics might also help the food issue.
- Governments are trying to diversify their economies to cope with falling demand for oil and gas.

## Key vocabulary

absolute monarchs, desalination, cloud-seeding, hydroponics, diversify

## Skills practised in 'Your turn'

- Geographical skills: q1, interpret a map
- Literacy skills: q2, define
- Thinking skills: q2, q4, q5, q6, explain; q3, q8, make a choice and explain it; q4, give reasons; q5, suggest reasons; q7, give an example

## Unit outcomes

By the end of this unit most students should be able to:

- explain the terms given in 'Key vocabulary' above
- name the countries of the Arabian Peninsula
- explain why water levels in the aquifer below the Arabian Peninsula are falling
- explain why Saudi Arabia has invested in farmland overseas
- give an example of how some governments in the Arabian Peninsula are trying to diversify their economies

## Ideas for a starter

1. With books closed, say: The Arabian Peninsula has seven countries. Who can name them? And now tell me their capitals (and mark them on a blank map).
2. With books closed, say: You are an astronaut orbiting Earth in the International Space Station. How will you recognise the Arabian Peninsula? (Display an image of the Arabian Peninsula from space.) What challenges do you think the area might be facing?

## Ideas for plenaries

1. Ask: How does having oil and gas reserves make a country wealthy?
2. What if … no oil or gas had been discovered in the Arabian Peninsula?

## Further class and homework opportunities

Suggestions 16–18 on page 163 of this book

*geog.3 workbook*, page 61

**kerboodle**

### Resources

7.5 You decide! – Students rank the potential solutions to the water shortage.

7.5 Word scramble – Students complete a crossword of key terms about environmental challenges.

7.5 Foundation worksheet – Students recall two key facts about the Arabian Peninsula and write a short article explaining one challenge the Peninsula faces.

7.5 Extension worksheet – Students write a short article about one challenge the Arabian Peninsula faces.

### Assessment

7.5 End-of-lesson test – An assignable, auto-marked formative assessment.

## Answers to 'Your turn'

1 Kuwait, Bahrain, Qatar, UAE, Oman, Saudi Arabia, Yemen

2 **a** A monarch who has absolute power over their people; for example makes all the country's laws.

**b** No, in the UK the monarch is a *constitutional* monarch; a head of state with no real power.

3 Answers will vary but all are valid, if they are justified with a strong argument.

4 Water levels in the aquifer are falling because it is being used by the people that live in the region. It is pumped out of the aquifer for domestic use, but also for industry and agriculture. The price does not reflect its scarcity, so people use too much. It is not being replenished as quickly as it is being used, because the region has little rain. (The aquifer contains fossil water and was formed about a million years ago).

5 **a** Saudi Arabia has invested in land in Sudan so that it can grow crops, which it can import to feed its population. The climate in Sudan is more suitable for farming.

**b** Some people in Sudan may be unhappy about this because it means they will have less fertile farmland to grow crops for their own people. The rights of people who live in these regions may be ignored. If dams are constructed for collecting water, other places in Sudan might suffer droughts.

6 Hydroponics are more sustainable than planting crops in the desert because water and nutrients can be recycled via the greenhouse system. The traditional systems of agriculture in these regions are very wasteful, with water and nutrients being lost into the ground. Hydroponics systems are also using solar power to desalinate sea water, which will be useful in the future if the fresh water runs out.

7 To 'diversify' means to widen or vary the scope of something. Countries can vary their economies by producing a wider range of exports that they can sell, or services that they can provide. Saudi Arabia sells oil and gas, but it is also promoting tourism.

8 A shortage of water could make it difficult for the Arabian Peninsula to diversify because the new ventures are likely to require large amounts of water. Culture, sports and shopping malls will attract more people, to live, work, and enjoy holidays in the Middle East. On the other hand, using desalinated sea water is an option. If governments invest in sustainable technology, the risks may be reduced. However, this technology is expensive and could slow down their plans to diversify.

# Conflict in the Middle East



## About this unit

This unit provides an overview of some of the conflicts in the Middle East.

## Key points

- The Middle East has long been an unstable region. Page 135 of the students' book shows some of the main conflicts (in 2021).
- Reasons for the instability include: the involvement of foreign countries in the region (now and in the past); tension between Sunni and Shia Muslims; the operation of rebel groups including IS and al-Qaeda.
- Several countries have been involved in most of the main conflicts in the Middle East.
- When a country is affected by conflict, development is reversed; ordinary people and children suffer.

## Key vocabulary

Sunni, Shia, Islamic extremists, Islamic State (IS), al-Qaeda, geopolitics, sanctions

## Skills practised in 'Your turn'

- Geographical skills: q1, q4, interpret a map; q6, q7, interpret a graph; q6, interpret a table; q6, compare sources and draw links
- Literacy skills: q2, define
- Thinking skills: q3, outline reasons; q3, q5, explain; q4, suggest a reason; q6, give evidence; q7, suggest a reason

## Unit outcomes

By the end of this unit most students should be able to:

- explain the terms given in 'Key vocabulary' above
- outline four reasons why foreign countries get involved in the Middle East
- explain why Turkey and Lebanon have high numbers of refugees
- explain how conflict can affect a country's development

## Ideas for a starter

1. With books closed show students the image of the children from page 134 of the students' book. Explain that they are refugees from Syria, living in Turkey. They could be your siblings. Ask: How does conflict affect ordinary people like these children?
2. Ask several students to read out information from some of the text boxes on page 135 of the students' book. Then ask the class: Why do you think there is conflict in the Middle East? Draw out ideas, and record them on a spider map.

## Ideas for plenaries

1. Ask: Do the conflicts in the Middle East affect the UK and the rest of the world? How?
2. What is geopolitics?
3. What has conflict got to do with geography? (Hint: geopolitics.) Discuss.

## Further class and homework opportunities

Suggestions 19–21 on page 163 of this book

*geog.3 workbook*, page 62

kerboodle

### Resources

7.6 You decide! – Students rank the potential effects of conflicts.

7.6 Sort it out! – Students sort the features of conflict in the Middle East.

7.6 Foundation worksheet – Students explain how five countries in the Middle East have been affected by conflict.

7.6 Extension worksheet – Students analyse how seven countries in the Middle East have been affected by conflict.

### Assessment

7.6 End-of-lesson test – An assignable, auto-marked formative assessment.

## Answers to 'Your turn'

**1** **a** Sunni **b** Shia **c** Sunni **d** Sunni

**2** **a** Extremists are people with extreme political or religious views; some are willing to take violent action in support of their views.

**b** Geopolitics is how human and geographical factors shape the relationships between countries; for example a country may want to stay friendly with a country that borders it, or has resources it needs.

**c** Sanctions are penalties a country places on another country, to punish it; for example it may refuse to buy goods from that country.

**3** **a** For a supply of oil and gas; to prevent the spread of extremism; to sell things (including weapons); to gain political influence.

**b** Countries want influence in other countries because it gives them power and access to resources.

**c** Two reasons are connected to trade: for a supply of oil and gas; to sell things (including weapons)

**4** Lebanon and Turkey have many refugees as a result of the conflict in the neighbouring country of Syria.

**5** Young people in Yemen may have been affected by the civil war because some may have been recruited as soldiers; some civilians have lost their lives due to the violence and many have died as a result of the famine caused by the war. As the country's development has been so badly affected, young people in Yemen are faced with a future of extreme poverty, without opportunities for education.

**6** **a** **i** D shows that Iran's GNI is about $13 000 per year. This makes it one of the poorer countries in the region, which is surprising since Figure E shows that Iran is #1 for gas reserves and #2 for oil reserves. We might expect its GNI to be higher considering its resources, so this is evidence that sanctions have affected its economy.

**ii** Iraq is in a similar situation, with a GNI of $11 000 a year putting it in the bottom five countries in the Middle East in terms of wealth; yet it is in the top five for both oil and gas reserves. This is evidence that its economy has suffered from war.

**b** These values have been estimated because the data was not available. Yemen and Syria were engaged in conflict at the time, which made it too difficult for their governments to collect data.

# The Middle East



## Skills practised in check questions

- Geographical skills: q1, q2, q3, interpret and compare maps; q2, interpret a photo; q3, interpret a table; q4, use a map to provide evidence
- Numeracy skills: q3, divide
- Literacy skills: q1, draw and complete a table; q2, define; q7, extended writing
- Thinking skills: q1, q3–q6, explain; q2, q3, suggest reasons

## Answers for check questions

**1 a**

| Country | Label on map A | Capital city |
|---|---|---|
| Egypt | f | Cairo |
| Saudi Arabia | h | Riyadh |
| Turkey | b | Ankara |
| Yemen | j | Sanaa |
| Iraq | d | Baghdad |
| Iran | e | Tehran |
| Oman | k | Muscat |

**b i** g is the Red Sea. a is the Mediterranean Sea.

**ii** Under the Red Sea, the Arabian plate is moving away from the African plate. This is a constructive plate boundary where volcanoes are can form.

**c** Two from: Syria – l; Lebanon (no letter); Yemen – j

**d** Qatar – m

**2 a** m

**b i** Narrow sea passages which ships pass through.

**ii** i – Strait of Hormuz; n – Bab-el-Mandeb Strait; c – Suez Canal

**c i** Egypt

**ii** It is a shortcut to the Mediterranean Sea. The only other ways ships could get there from the Middle East or western Asia is by sailing around the whole of Africa, so many are willing to pay the high price for passage. Perishable goods would not survive the longer journey and more fuel would be required.

**d** The UK may stop buying LPG (liquefied petroleum gas) in the future because it comes from oil. The use of oil as a fossil fuel is contributing to the climate crisis. The UK aims to be carbon neutral by 2050.

**3 a i** Yemen is the most densely populated. The population of the countries is similar, but Yemen is a much smaller country than Saudi Arabia in terms of its area.

**ii** About 47 times

**iii** Two from: Saudi Arabia has more oil and gas reserves than Yemen; Yemen has suffered years of conflict and has been in a civil war since 2014; Saudi Arabia has also started to make money from tourism.

**4 a** Farmers can get fresh water from the huge aquifer system which lies under the Arabian Peninsula.

**b** This may not always be available because it is being depleted faster than it is being replenished.

**c** Desalination

**d** Turkey. Map B on page 128 provides evidence for this. Some of Turkey has a Mediterranean climate which experiences warm, wet winters.

**5 a** Benefits: oil and gas resources have made the Middle East an important region and give it political power; selling oil and gas has made some countries very wealthy; other valuable products can also be manufactured from oil and gas (eg plastics and fertilisers).

**b** Instability: oil and gas wealth has led to great inequality in the region; foreign countries have intervened in Middle Eastern politics to ensure they have access to oil and gas, which has created conflict; conflict reverses development.

**6 a** Demand for oil and gas will decline because climate change has forced world leaders to make plans to increase the use of green energy and reduce the use of fossil fuels.

**b** The countries of the Arabian Peninsula will find this especially challenging because exports of oil and gas dominate their economy. Some countries have other valuable exports such as pharmaceuticals and hi-tech equipment, but these sectors are still quite small.

**7** For this extension question, students are asked to state the extent to which they agree with a statement. Good answers will give evidence to support both sides of the argument and provide a short conclusion.

An example answer: Studying the Middle East has demonstrated some ways in which the world is interconnected. Oil and gas which is extracted in the Arabian Peninsula fuels many richer countries; and conflict in the Middle East can be linked to intervention from other countries such as the UK and USA.

However, there are other aspects of geography which could be studied to help understand the global 'big picture' of interconnectedness. Learning about the impacts of climate change in the Arctic would show the problems created by burning fossil fuels and deforestation; and studying the newly industrialised countries in Asia would shed light on world trade.

Overall, learning about the Middle East can demonstrate interconnectedness to some extent, but there is always more to learn.

# Further suggestions for class and homework

Most of these suggestions are addressed to your students. Where research or further resources are needed, the Internet will almost certainly provide the answer.

The suggestions are graded *, **, *** according to level of difficulty. Some are suitable for all levels, and can be differentiated by outcome.

## Introducing the Middle East

1 **Colour in** On a blank map of the Middle East, colour in and label the countries. Mark in the Tropic of Cancer. Label the ocean / gulfs / straits / seas, and the Suez Canal. (You'll need the map on page 126 of *geog.3* for some of this.) */**

2 **Snap** Make a pack of cards, with names of the Middle East countries on half of the cards, and their capitals on the other half. Devise a way to play Snap! Or a memory game. **/***

3 **How big?** Find the area in sq km of each Middle East country, and the UK (try the Internet). Show the areas on a bar graph, with vertical bars. In order, largest first. Use one colour for Middle East countries and a second colour for the UK. Make up a suitable title. ***

4 **Adopt a Middle East country and ...** write a unit about it, like the units in *geog.3* – but for ten-year-olds. Give at least six geographical facts about it, and include a map. Add photos too, with captions. (This could become part of a wall frieze, covering all the countries. Students work in pairs?) */**/***

## The Middle East: physical geography

5 **Word search** Create a word search for the countries of the Middle East, their capitals, and physical features. Try it out on a partner. You can write the words in any direction. */**/***

6 **Your oil tanker** You are arranging a shipment of oil from Kuwait to an oil refinery in Southampton, in the UK. Your tanker will pass through the Suez Canal. You have to mark its route on a map, naming all the seas, gulfs and straits it will pass through, and all the countries it will pass by. Make sure the map is big enough for all this. You could make up a name for your tanker and include a photo of it. **/***

7 **The Suez Canal** It has someone in charge – and that's you! Tell us about the canal – which country owns it, how long it is, how busy it gets, why it's so good for that country, and others countries too. ***

8 **Why does the Red Sea have active volcanoes?** Carry out an enquiry. (You might need to help students find a suitable plate map.) Include a photo of its volcanic islands. ***

## The Middle East: climate zones and biomes

9 **Climate graphs** Look in Google Images for climate graphs for capital cities in the Middle East, plus a city in the UK. Choose graphs with a single average temperature line (not max and min). Give pairs of graphs to pairs of students to compare. Which is hotter in summer? Why? Which gets cooler? Why? Which gets more rain? **/***

10 **What's the weather like?** Choose a city in the Middle East. Look up its weather forecast for today. Add the country name in the search box. (Try a site such as *www.weather.com*.) What's the temperature? Precipitation? Wind speed? Compare it to the weather forecast for your nearest city. What differences are there? Explain them. **/***

11 **Visit the pyramids!** Plan a trip to the pyramids in Egypt. How will you get there? What will you see? Any plants or animals? Will it be dry, hot, wet, cold? Explain. **/***

## The people, and economies

12 **Populations** Page 130 has a table of populations. Show these on a bar graph, with vertical bars. In order, largest first. Use two colours, one for Middle East countries and the other for the UK and London. Give your graph a title. **/***

13 **Needing the Nile** You are a geography student in Egypt. Explain why the Nile is so important to your country, and how people use its water. For example your neighbour, who is a farmer. How would you like to present this information? Don't forget a map of the Nile valley and delta. **/***

14 **Middle East food** Find a recipe you like the look of, from a Middle Eastern country. Follow it! And share? **/***

15 **Writing** Show the class examples of writing in Arabic, Turkish, Kurdish, Persian (Farsi), Hebrew. */**

## Challenges in the Arabian Peninsula

16 **Life beyond oil and gas** Demand is falling, and you are in charge of plans to diversify the economy for one of the oil and gas producing countries of the Middle East (check the table on page 131). Choose a country and formulate your plan. Work in pairs. */**/***

17 **Alphabet run** Try a complete alphabet run for the Middle East using what you've learned. An atlas might help. **/***

18 **Food risk? Water stress?** Choose one of these challenges. Devise an enquiry question for the class to undertake. ***

## Conflict in the Middle East

19 **The Middle East in the news** Draw up a table with two columns, the names of the Middle East countries in one. The other is for frequency of mention. Record how often each Middle East country is mentioned in the news / newspapers, daily for at least two weeks, with a tally mark for each mention. Use two different colours for your tally marks: one for news of conflict, the other for all other mentions. Discuss the findings. (You could allocate this as a rolling task to different students.) **/***

20 **Explain it!** You have a visitor from another solar system – who does not understand why Earthlings can't live in peace and harmony. Make up a conversation between the two of you. (This visitor can speak perfect English!) **/***

21 **Give peace a chance** You want to make a difference, and start a worldwide peace movement. How will you go about it? (Do you think peace starts in your local area?) */**/***

# Glossary

## A

**absolute monarch** – a monarch who has complete control over his or her people

**add value** – process something in a way that increases its value; so you can charge more

**adult literacy rate** – the % of the population aged 15 and over who can read and write a simple sentence about everyday life

**altitude** – height of a place above sea level

**aquifer** – a natural underground rock structure which holds groundwater

**area of deprivation** – an area where people suffer disdvantages such as high unemployment, poor health, poor housing

**arid** – very dry; receives little or no rain

**artificial intelligence** – when machines are programmed to do tasks that usually need human intelligence

## B

**bedrock** – the solid rock underlying the soil

**biodiversity** – the variety of living things in a place

**biological weathering** – the breaking up of rock by plant roots and burrowing animals

**biomass** – plant or animal material used as fuel; for example wood, and waste from farms

**biome** – a large area with a similar climate, plants, and animals

## C

**carbon neutral** – adding no carbon dioxide to the atmosphere, overall; if any is being added, it is cancelled out by planting trees

**cash crop** – a crop you grow for sale

**chain store** – one of a group of similar shops that are owned by the same company

**chemical weathering** – rock is broken down by chemical reactions, for example with rainwater

**climate change** – all aspects of climate are changing because Earth is getting warmer

**climate zone** – a large area with roughly the same climate throughout (so it will have the same vegetation and animals too)

**commodities** – agricultural produce and natural materials, sold in bulk; for example coffee, iron ore, oil

**complex volcano** – a landform made up of several related volcanoes

**conflict** – serious disagreement, which may lead to violence and even war

**coniferous** – describes trees which bear cones (such as pine trees)

**corruption** – dishonest conduct by people in official positions; for example, they may accept bribes

**crater** – the hollow around a vent in a volcano

## D

**deciduous** – describes trees which lose their leaves in winter; for example oak trees

**decline** – to fall gradually into a poor state

**deposit** – to drop material; waves deposit sand and small stones to form beaches

**desalination plant** – where seawater is turned into fresh water, by removing its salt

**desert** – gets very little rain; it can be a hot or cold desert, and sandy or rocky

**developing country** – its people are poor, on average, and lack many services

**development** – a process of change to improve people's lives

**development indicator** – a piece of data that helps to show how developed a country is

**diversify** – develop a wider range; for example, more ways to earn income

## E

**earthquake** – the shaking of Earth's crust, caused by sudden rock movement

**economic** – about money, jobs, and business

**economic migrant** – a person who moves in order to find paid work, or to earn more

**economy** – all the activities going on in a country, in producing, buying, selling, and distributing goods and services

**emerging economy** – its development is speeding up, usually because of industry

**employment** – the state of having paid work

**employment structure** – the percentage of the workforce in each sector: primary, secondary, and tertiary

**environment** – everything around you; air, soil, water, animals, and plants form the natural environment

**erosion** – the wearing away of rock, stones and soil by rain, rivers, waves, wind, or glaciers

**exclave** – part of a country that is cut off from the main part; you pass through another country (or countries) to reach it

**export** – sell things to other countries

**extreme poverty** – where people have less than $1.90 a day to live on; this figure is set by the World Bank, and can change

**extremists** – people with extreme political or religious views; some are willing to take violent action in support of their views

## F

**fault** – a crack in rock, where blocks of rock can move relative to each other

**fertile (soil)** – able to produce healthy crops

**fertilisers** – substances added to soil to make it more fertile

**finite resource** – there is a limited amount of it (so it could run out one day)

**fold mountain** – formed by plates pushing into each other; the rock at the plate edges gets folded upwards, making mountains

**food bank** – a place which gives out free (donated) food to people who need it

**food insecurity** – people do not have access at all times to the food they need (see next)

**food security** – people have access, at all times, to enough safe and nutritious food for a healthy and active life

**fossil fuels** – coal, oil, natural gas

**fossil water** – groundwater that has been sealed into aquifers for thousands or millions of years, by changes in the surrounding rock

**fresh water** – water that is not salty; for example the water in rivers and springs

**fumarole** – a vent or opening in Earth's crust which emits steam and gases

## G

**geopolitics** – how human and geographical factors shape the relationships between countries; for example a country may want to stay friendly with a country that borders it, or has resources it needs

**geyser** – a spring of water that shoots into the air every so often; it is heated by hot rock below ground, and the pressure builds up

**globalisation** – how the world is becoming more interconnected, through movement of goods, people, money, and information

**GNI (gross national income)** – the total amount that a country's population and businesses earn in a year

**GNI per person** – the GNI divided by the population; it is a measure of how wealthy the people in a country are, on average

**GNI per person (PPP)** – the GNI per person is adjusted to take into account that things cost more in some places than others

**goods** – physical objects, such as pens, apples, and cars, that are bought and sold

**green electricity** – generated using renewable resources such as wind, waves, sunlight; no harmful emissions are produced

**groundwater** – water that collects below ground, when rain trickles through the soil

## H

**habitat** – the natural environment of a species; its home

**human development index (HDI)** – a score between 0 and 1 that indicates how developed a country is; it combines data on life expectancy, education, and income

**hydroelectricity** – electricity generated when flowing water spins a turbine

**hydroponics** – growing plants in water that contains the nutrients they need

## I

**igneous rock** – forms when melted rock hardens

**import** – buy in things from other countries

**Industrial Revolution – the period (about 1760** – 1840) when many new machines were invented in the UK, and factories built

**inequality** – when wealth and access to services are not shared equally

**infrastructure** – facilities such as roads, water supply, electricity grid, and railways, that keep a country and its economy running

**irrigate** – to water crops

## L

**landlocked** – surrounded by land, with no coastline

**latitude** – how far a place is north or south of the Equator; it is measured in degrees

**latrine** – a very basic toilet; it could be just a hole, or trench, in the ground

**lava** – melted rock at Earth's surface

**life expectancy** – how many years a new baby can expect to live for, on average

**living sustainably** – living in a way that does not harm us humans, other species, or the environment

**longitude** – how far a place is east or west of the Prime Meridian; it is measured in degrees

## M

**magma** – melted rock below Earth's surface

**magnitude** – how much energy an earthquake gives out (measured on the Richter scale)

**manufacturing** – making things in factories

**mass extinction** – when a large number of species die off over quite a short period of time; for example because of an ice age

**metamorphic rock** – forms when rock is changed through the action of heat and / or pressure, without melting

**migrant** – a person who moves somewhere else, usually to find work, or a better life

**mineral** – a natural compound in rock

## N

**natural resource** – it occurs naturally in the environment, and we can make use of it; for example, wind and oil

**NGO (non-governmental organisation)** – a charity that helps people, and is not linked to a government; for example, Oxfam

**non-renewable resource** – a resource that is limited, and could run out one day

**North Atlantic Drift** – a warm current in the Atlantic Ocean; it flows up to the Arctic

**nutrients (for plants)** – substances which plants need, to grow and be healthy

## O

**out-of-town shopping centre** – a centre built for shopping, outside a town or city, with a range of shops, and free car parking

## P

**pandemic** – a disease that spreads over a very wide area, or the whole world, and affects a large number of people

**peninsula** – land that juts out into the sea, and is almost surrounded by water

**permafrost** – the ground under the surface that is permanently frozen, in the tundra

**permit** – a document giving a person permission to do something

**physical weathering** – weathering which breaks rock into smaller bits

**plateau** – an area of fairly flat high land

**plates** – Earth's hard outer part is broken into big slabs called plates, which move around

**populous** – has a large population

**poverty** – the state of being poor

**poverty line** – where a sum of money is used to define the level of poverty in a country; for example, the % of people living on less than \$1.90, or \$5.50, a day

**precipitation** – water falling from the sky (as rain, sleet, hail, snow)

**primary products** – goods like oil and timber, from the primary sector of the economy

**primary sector** – the sector of the economy where people collect things from Earth; farmers and miners are in this sector

**PV cell (photovoltaic cell)** – converts the energy from sunlight into electricity

## Q

**quality of life** – the level of comfort and well-being a person enjoys

**quaternary sector** – the sector of the economy where people use high-level expertise to develop things that will help other sectors

## R

**refugee** – a person who has been forced to flee from danger; for example from war

**relief** – how the height of the land varies

**renewable resource** – a resource we can make use of, that will not run out; for example, sunlight and wind

**replacement fertility rate** – the average number of children per woman that keeps a population at the same size (without migration); it is taken as 2.1 children

**republic** – does not have a monarch

**resources** – things we need to live, or use to earn a living; for example food, fuel

**Ring of Fire** – the ring of volcanoes and earthquake sites circling the Pacific Ocean

## S

**sanctions** – penalties a country places on another country, to punish it; for example it may refuse to buy goods from that country

**sedimentary rock** – formed from sediment; sandstone is formed from a sediment of sand

**secondary sector** – the sector of the economy where people build things and make things, for example in factories

**services** – activities carried out to meet needs, such as teaching, or treating tooth decay

**social** – about people and society

**solar power** – electricity generated from sunlight, using photovoltaic (PV) cells

**standard of living** – the level of goods, services, and comfort available to people

**steppe** – a large flat area of treeless grassland

**sustainable** – can be carried on into the future without doing harm

## T

**taiga** – region of coniferous forests which lies between the tundra and steppe

**tariffs** – taxes on imports or exports

**temperate** – term used for a mild climate

**tertiary sector** – the sector of the economy where people provide services for other people; for example healthcare

**trade war** – where countries try to damage each other's trade; for example, they may place tariffs on imports from each other

**transnational corporation (TNC)** – a company that operates in more than one country; examples are Apple, Nike, Nissan

**tree line** – the line or altitude above which it is too cold for trees to grow

**tsunami** – waves generated by an earthquake in the ocean floor

**tundra** – a cold region where the ground is deeply frozen; only the surface thaws in summer, allowing small plants to grow

## U

**under-5 mortality rate** – the % of babies born alive who die before age 5

**unemployment** – when people are looking for paid work, and can't find it

**utilities** – services provided, such as electricity, gas, and water supplies, sewage removal, and phone lines

## V

**volcano** – where lava erupts at Earth's surface

## W

**water stress** – where a country or area cannot meet its demand for fresh water

**weathering** – the breaking down of rock, by the action of things in its environment: heat, cold, rain, plant roots, and so on

# Command words glossary

Throughout *geog.123* students will meet questions with **command words**.

Command words in questions tell students what they need to do. It's important that students understand the command words in order to answer questions fully and well.

And if students can answer questions well, they're likely to enjoy their course more!

Later, students will meet command words in exams, so getting used to them, and understanding how to respond to them, will prove very beneficial.

Here are the command words used in *geog.123*, in alphabetical order – along with their definitions and a short summary of how to respond to the command word.

## Assess

**Weigh everything up and make a judgement.**

For example assess the impact of an earthquake on a city. Answers should always include the evidence that a judgement is based on.

## Calculate

**Do some maths, to get the answer!**

Answers should always include the unit. For example *5 km* or *11 people* or *15 days*. (Not just *5* or *11* or *15*.)

## Compare

**Identify what is the same, and different, about two things.**

For example say which one is bigger. Answers should always mention *both* things.

## Copy and complete

**Copy this, filling in all the blanks.**

Fill in using the words and terms that make sense!

## Define

**Write down the meaning.**

Answers should be kept clear and simple.

## Describe

**Write a description.**

For example describe what can be seen, or the steps in a process. It is not necessary to give reasons for anything.

## Discuss

**Look at something from different angles, and give key points about it.**

For example give its good and bad points, or its benefits and drawbacks.

## Draw

**Draw!**

For example draw a diagram, or sketch map, or bar chart, or line graph. Remind students to use a ruler for straight lines. And that accuracy is important for bar charts and graphs. Encourage students to be quick!

## Evaluate

**Judge how successful or worthwhile something is.**

Answers should say what has been good and bad about it, *and* give a final opinion.

## Examine

**Look at each part and say how it contributes.**

For example examine how different processes work together to form an oxbow lake.

## Explain

**Make something clear and easy to understand.**

For example explain how a meander forms.

## Give

**Come up with an answer, from what you've learned.**

Remind students to keep it clear and simple.

## Identify

**Pick out the thing, and give its name.**

For example identify a landform on a map.

## Justify

**Give reasons to support the choice or decision you made.**

For example give reasons why you agree with a statement.

## Label

**Add labels!**

For example label a diagram. Remind students that the aim is to make it clear and easy to understand – so encourage them to keep labels short and simple!

## Name

**Write the name of the thing you are asked about.**

Answers do not need a full sentence.

## Outline

**Set out the main points.**

Stick to the main points. Answers don't need to give details.

## State

**Give the answer in clear terms.**

*State* is often used in place of *Give* or *Identify* or even *Calculate* or *Count*. Remind students about the importance of answering clearly.

## Suggest

**Come up with a possible reason or plan.**

Remind students to use their common sense!

## To what extent .... ?

**How much does it contribute, or how important / true is it?**

Answers should make a judgement, and give the evidence it's based on.

# Engagement, rigour, and progression
## delivered by a trusted series

These new editions provide the solid grounding students need for GCSE study. Retaining the motivating approach that has made it the number one course for ages 11-14, the 5th edition has been thoroughly updated to prepare students for the increased rigour and content of GCSE Geography. Kerboodle provides time-saving assessment, homework materials, and support for non-specialists.

## What's in geog.123?

| | STUDENT RESOURCES | | | TEACHER SUPPORT | |
|---|---|---|---|---|---|
| GEOG.1 |   geog.1 Student Book 978 019 844604 0 |   geog.1 Kerboodle Book (Student access) 978 019 839314 6 |   geog.1 Workbook 978 019 844606 4<br>Workbook Answer Book 978 019 844607 1 |   geog.1 Kerboodle Lessons, Resources and Assessment 978 019 839311 5 |   geog.1 Teacher Handbook 978 019 844608 8 |
| GEOG.2 |   geog.2 Student Book 978 019 848915 3 |   geog.2 Kerboodle Book (Student access) 978 019 839315 3 |   geog.2 Workbook 978 019 848986 3<br>Workbook Answer Book 978 019 848987 0 |   geog.2 Kerboodle Lessons, Resources and Assessment 978 019 839312 2 |   geog.2 Teacher Handbook 978 019 848988 7 |
| GEOG.3 |   geog.3 Student Book 978 019 848991 7 |   geog.3 Kerboodle Book (Student access) 978 019 839316 0 |   geog.3 Workbook 978 019 848993 1<br>Workbook Answer Book 978 019 848994 8 |   geog.3 Kerboodle Lessons, Resources and Assessment 978 019 839313 9 |   geog.3 Teacher Handbook 978 019 848995 5 |